Language and Information

No. 28

Bampton Lectures in America
Delivered at Columbia University

Bampton Lectures in America

1948 Arnold J. Toynbee: *The Prospects of Western Civilization*
1949 Paul R. Hawley: *New Discoveries and Their Effect*
1950 Charles H. Dodd: *Faith and Ethics in Early Christianity*
1951 Lewis Mumford: *Art and Technics*
1952 James B. Conant: *Modern Science and Modern Man*
1953 Alan Gregg: *Where Medicine Belongs Today*
1954 John Baillie: *The Idea of Revelation in the Light of Recent Discussion*
1955 Lionello Venturi: *Four Steps Toward Modern Art*
1956 Joel H. Hildebrand: *Science in the Making*
1957 Brock Chisholm: *The Expanding Concept of Health*
1958 Eric Lionel Mascall: *The Importance of Being Human*
1959 Sir Anthony Frederick Blunt: *The Art of William Blake*
1960 Detlev W. Bronk: *The Status of Science in Modern Society*
1961 W. Barry Wood, Jr.: *From Miasmas to Molecules*
1962 Paul Tillich: *Christianity and the Encounter of the World Religions*
1963 Northrop Frye: *The Development of Shakespearean Romance*
1964 Fred Hoyle: *Man and the Universe*
1965 Robert Hanna Felix: *Mental Illness: A Yielding Enigma*
1966 Alasdair MacIntyre: *The Dispute about God: Victorian Relevance and Contemporary Irrelevance*
Paul Ricoeur: *Religion, Atheism and Faith*
1968 Sir John Summerson: *Victorian Architecture: Four Studies in Evaluation*
1969 Jacob Bronowski: *Magic, Science and Civilization*
1975 Paul Ramsey: *Ethical Issues in Modern Medicine*
1976 Symposium: *Titian—His World and His Legacy*
1980 Symposium: *Bernini and the Baroque*
1982 Anthony Kenny: *Faith and Reason*
1983 Steven Weinberg: *On the Art of Science*
1984 William Arrowsmith: *Innovation and Tradition in Euripides*
1986 Zellig Harris: *Language and Information*

Language and Information

Zellig Harris

Columbia University Press
New York
1988

Columbia University Press
New York Guildford, Surrey

Library of Congress Cataloging-in-Publication Data
Harris, Zellig Sabbettai, 1909-
Language and information.

(Bampton lectures in America; no. 28)
Includes index.
1. Linguistics. 2. Language and languages.
I. Title. II. Series.
P121.H34 1988 410 87-22400
ISBN 0-231-06662-7

Printed in the United States of America

Hardback editions of Columbia University Press books are Smyth-sewn and are printed on permanent and durable acid-free paper

Contents

Foreword

THE BAMPTON Lectures of 1986 present a formal theory of language structure, together with several additional results that follow from it. The first lecture summarizes methods of language analysis and a theory of language structure, which are formal in that the entities are defined by their frequency of occurrence relative to each other, rather than by phonetic or semantic properties. The relations and operations of the system are then defined as mappings on sets of these entities. The theory is not an invented model, but arises out of a search for regularities in the data of various languages. In application to English, it has been shown to yield a detailed grammar directly from first principles. Each entity and relation makes a fixed contribution to the meaning of its sentence; this meaning is thus obtained directly from the words and structure of the sentence.

The second lecture deals with the languages of science. This issue arises because the methods presented in the first lecture, when applied to the reports of a subscience alone, yield by virtue of their combinatorial basis a sublanguage grammar different from that of the whole language in which the reports were written. Furthermore, for a given subscience, the reports and discussions written in one language satisfy much the same special grammar as do papers in the same field written in other languages. The structure of each science language is found to conform to the information in that science rather than to the grammar of the whole language. In important features, the science languages are between natural languages and mathematics.

The third lecture is a discussion of information, in particular as it appears in language. This is not an added or independent issue, but an intrinsic concomitant of the structural analysis. True, the theory presented in the first lecture builds only upon the combinations, of sounds and of words, that constitute the utterances of a language. However, when the analysis is carried out in these terms, it is found that every step in the construction of a sentence has an established and largely unchanging contribution to the meaning of every sentence in which it occurs. There is thus an inherent relation between form and information in language. It is this that enables the analysis of the structure of a sentence in terms of the present theory to yield also its information. Furthermore, important properties of the structure of information can be garnered from the structure of language.

The last lecture considers what the structural theory presented here contributes to our understanding of the nature and development of language. The constructions presented in the theory are defined each on the basis of prior constructions. This creates a definitional order of constructions, which has a certain relation to the order of development in language. This developmental picture, together with the ongoing structural processes, suggest that the structure has grown from conventionalizations of use in speech, especially of word combinings. The processes that create and develop language and its informational capacity are found to have properties of a self-organizing and evolving system, and one geared to the transmission of information.[1]

There are, of course, other aspects of language than those discussed here. For example, there are varied uses and changes of language. Some affect the structure; others remain simply physical or cultural properties of language, possibly with social, geographic, or historical correlates. There are also variations made on grammar (chiefly on domains of reduction) and on word combination (in effect, on likelihoods) for purposes of play, and above all for purposes of art.

1. The methods and data on which these lectures are based cannot be given here. They may be found in previous work by the present writer, chiefly *(Methods in) Structural Linguistics* (Chicago: University of Chicago Press, 1951); *Mathematical Structures of Language,* Interscience Tracts in Pure and Applied Mathematics (New York: Wiley, 1968); *Papers in Structural and Transformational Linguistics* (Dordrecht: Reidel, 1970); and *A Grammar of English on Mathematical Principles* (New York: Wiley-Interscience, 1982).

Language and Information

1

A Formal Theory of Syntax

1.1 Problems and Methods

TO CONSIDER the structure of language, especially its syntax—that is, how sentences are built from words—we note first the usual approaches. Scientists coming to the problem from the outside often seek regularities in the sequential relation among the words in a sentence, since the data presents itself sequentially. However, sufficient regularities have not been found, for reasons that will appear later. In contrast, people who work with language analyze it on the basis of what is called grammatical relations, such as the subject and object of a verb, or the relation of an affix to its host word. Here too there are difficulties. Few if any grammatical relations appear in the same way in all languages, so the individual relations cannot be taken as primitives for language as such. The situation is rather that in each language there are some relations that can be called grammatical, but a satisfactory general definition is lacking. Furthermore, grammatical relations are unique to natural language, and if we can describe language only in such terms we will be unable to compare language with anything else in the world, not even with such close relatives as gesture on the one hand and mathematics on the other. Finally, the elements on which grammatical relations hold are not adequately defined. The one type of element that is precisely established is the set of phonemes, the characteristic sounds of a language; indeed the discovery of phonemes is the beginning of a precise science of language.[1] But as to words, if they are thought of as correlations of sound sequences with meanings, we are left with many problems, such as homonyms (as in *see, sea,* and the Holy *See*), or with the two pronunciations of *economics,* not to mention many exceptional situations. And as to sentences, the lack of a general definition is well known. Hence, while traditional grammars can provide adequate descriptions of a language, they do not supply a framework for considering the structure of language in general.

The question has thus to be reconsidered from the beginning: How can one investigate the structure of language? In general, the investigation of a field, and the defining of its entities, is carried out in a metalanguage of this field, a language of broader informational capacity than the given field. This is clearly so in mathematics and logic, where the precision as to what is in the field enables

1. The major original works on phonemes are: F. de Saussure, *Cours de linguistique générale* (Paris, 1910); Edward Sapir, "Sound Patterns in Language," *Language* (1925), 1:37–51; and Leonard Bloomfield, *Language* (New York: Holt, 1933).

us to recognize that statements said about the field are not in it; that is, they do not have the structure of what is said in the field. But natural language has no external metalanguage. We canot describe the structure of natural language in some other kind of system, for any system in which we could identify the elements and meanings of a given language would have to have already the same essential structure of words and sentences as the language to be described. The grammar of English can be written in English or French or any other language, and the grammar of French in French or English or any other language, but not in mathematics, or in gestures, or in any nonlanguage system.

In the absence of an external metalanguage, the entities of each language can be identified only if the sounds, markers, or words of which they are composed do not occur randomly in utterances of the language. That is, the entitites can be recognized only if not all combinations occur, or are equally probable. This condition is indeed satisfied by languages. A necessary step, then, toward understanding language structure is to distinguish the combinations of elements that occur in the utterances of a language from those that do not: that is, to characterize their departures from randomness. This task entails an important demand: it calls for a least description, that is, for a characterization of the actually occurring combinations by means of the fewest and simplest entities and the fewest and simplest rules and conditions of their combination, and with no (or least) repetition. The reason for this demand is that every entity and rule, and every complexity and restriction of domains of a rule, states a departure from randomness in the language being described. Since what we have to describe is the restriction on combinations in the language, the description should not add restrictions of its own. If two descriptions, one more efficient than the other, characterize the same data, then the less efficient description must have overstated the actual restrictions in the language—by overstating and then withdrawing part, or by repeating a restriction, or whatever.

1.2 Procedures Yielding the Elements

To see how the description of language structure is achieved, we note first how the elements can be established.

First, it is possible to determine the phonemic distinctions in a language by a behavioral test that does not involve the specific meaning of words or the investigator's judgment of phonetic similarity. The test consists of one speaker of a language uttering, in random order, repetitions of two words (e.g., *see* and *sea,* or *hard* and *heart*) while the hearer (another speaker) judges which pronunciations are repetitions of which. The underlying fact of language is that in speech people can distinguish repetition, which they cannot do as explicitly with gestures or many other behaviors: in English speech, *see* and *sea* are repetitions, *hard* and *heart* are not. These discriminations create sound types, as against merely a scientist's aggregation of sound tokens, and the discriminations themselves constitute the discrete and definite ("phonemic") elements out of which everything else in language is constructed. Note that homonyms are not distinguished by this test: *see* and *sea,* or *heart* and *hart,* are repetitions of each other. Phonemes are obtained by the most economical way of collecting into one element phonemic distinctions having different environments, e.g., the *ph* pronunciation in *pin* and the *p* pronunciation in *spin.*

Next, it is possible to locate word boundaries within utterances of a language (and morpheme boundaries, e.g., affixes, within words) by a stochastic process—that is, a process that checks the $n + 1^{\text{th}}$ item given the first n items. To find the word boundaries in, say, *If he comes call me,* we take many utterances beginning with *i* and see how many different second phonemes (or letters) occur in them, then may utterances beginning with *if* and see how many different third phonemes appear, then many utterances beginning with *ifh* and ask how many different fourth phonemes, and so on. In this specific series we find more different third ones than either second or fourth. Each point at which the number of different possible next phonemes (or letters) peaks, i.e., at which the number is greater than immediately before or after, is (in most cases) a word or morpheme boundary. Such peaking arises because not all phoneme sequences make words and not all word sequences make sentences.

Finally, it is possible to locate the boundary of sentences within utterances. When sequences of words in utterances are studied, it is found that to a certain extent one can classify the local combinations into required ones (where words of a given class are required after a given word) and several kinds of possible (permitted) ones (for all the classes whose words can occur right after the given word). A complex stochastic process on the word sequence in an utterance, in respect to these types of successor classification, reveals a periodicity: at certain points, the successor possibilities return to the situation at the beginning of the utterance. These recurrent-event points segment the utterance into a succession of structurally independent sentences.

The stochastic processes just mentioned are important even for a known language, where we know from experience what are the words and the sentences. First, the processes show that words and sentences exist, not merely by cultural convention or by some semantic properties but by restrictions of combination, that is of occurrences relative to each other, in the physical components of speech. Secondly, they show that each type of entity is definable as a relation holding among entities of a smaller (more local) type. The phoneme sequence relation that makes words proves to be of little interest: it does not in general tie up with the other properties of words—not with their meanings (for reasons considered in the third lecture) and not with the word sequence relation that makes sentences. However, the word sequence relations are explicit and are of decisive importance for the structure and meaning of sentences. It will be seen that these last relations are primarily a matter of the frequency of words relative to each other in utterances, or sentences, of a language.

1.3 Syntax Procedures

Before investigating the relative frequencies, we must see how the data of language can be rearranged so as to reveal regularities in these frequencies.

We begin by recognizing a relation among sentences, which will later (1.6) be seen to form a set of partial transformations in the set of sentences. Roughly, this relation holds between two subsets of sentences in each of which there is approximately the same grading of acceptability (used here as an experimental intuitive measure) for particular word choices, but with a constant difference in form between the two subsets. The transformation is thus a function, where the difference in form operates on each sentence of one subset to yield ("derive") the corresponding (image) sentence in the other subset, with the sentence and its image turning out to be roughly paraphastic.

To take a simple example, consider the set of sentence segments under *and* where one or more words in the second sentence are the same as the correspondingly positioned words in the first: e.g., *I knocked and I entered; John came over and John introduced Mary; John came over and John introduced himself; Mary came over and Mary introduced herself.* We represent this set by *John V and John W*. To this corresponds a second set (represented by *John V and W*): e.g., *I knocked and entered; John came over and introduced Mary.* There are various ways of showing that the *John V and W* set can be derived from *John V and John W* by reducing to zero any word in the second sentence that repeats the same-positioned word in the first sentence. One way is to ask the hearer to supply any word that may be considered missing in, e.g., *I knocked and entered.* A more formal way is to show that the least grammar that accounts for *John V and W* goes via *John V and John W* plus the zeroing just stated. For this, we note that the *John V and W* set contains *John came over and introduced himself, Mary came over and introduced herself,* but not *John came over and introduced herself.* Since *John introduced himself* is a sentence but *John introduced herself* is not, we know that *himself* repeats the subject of the preceding verb. To explain *John came over and introduced himself* (but not *herself*) we would have to extend this condition and say that *himself* can also repeat the subject of the verb preceding the *and;* but this is a new statement for the grammar (and not as simple to formulate as it seems, because many words may intervene). If, however, we say that the missing subject of the second verb is a zeroed repetition of the first subject, then the second verb gets a subject to which the *himself* can refer, and the existing unextended restriction on *himself-herself* suffices for the second verb too. In addition, this zeroing fits all internal and external conditions on the sentence, such as the relation of the subject to each of the two verbs: the acceptability ordering of the various members of W (e.g., *spoke, slithered, melted* in *John V and W* is about the same as their ordering in *John W.*) We then say that *John V and W* is derived from *John V and John W* as its source.

Indeed, the great bulk of the sentences of a language are derived by transformation—to a large extent by reduction—from other sentences, their sources. We now consider all these source sentences, together with any other sentences that have not been shown so far to be derived from anything else. In thse nonderived sentences, we investigate the regularities of word occurrence, which turn out to be far more visible here than in the transformed sentences. Here we may consider first the shortest sentences, because it is found that the longer ones can be built up in a few regular ways from the shorter ones. In these short sentences it is found that certain words never appear in stated positions in respect to other words: for example, we may find *The children slept, The trees slept during the winter,* even *The stones slept (through the ages),* or *The universe will then sleep (until the next big bang),* but not *The is slept* or *The up slept.* Of words that do appear in the given position, some may be far more frequent (in a sufficiently large sample) than others. Even if we simply ask speakers of a language for estimated frequency, probability, or likelihood of words in stated short sentential environments, we find that some words will be excluded, e.g., *is* or *up* before *slept.*

We are now ready to consider what combinations of words occur in the language, in contrast with those that do not. This cannot be done by simply listing them. First, the list would be too vast. Second, the set of sentences is not well defined: there are many marginal sentences about which speakers are not certain or do not agree about whether they are said at all, or are in the language. Third, language changes, and no list would be correct over a sufficient period. Instead of listing, therefore, we try to find what constraints preclude the combinations that are not in the language, what restrictions affect the equiprobability of word occurrences in respect to each other in utterances of the language. It will be seen that there are three types of constraint on word combination that make sentences, and that each one carries a type of meaning, so that the meaning of a sentence is determined directly from the words and the constraints. The three are: a partial ordering that creates sentencehood, a probability inequality that allows for word meaning, and a reducing of phonemic forms that does not affect the objective meaning.

1.4 The Partial-Order Constraint

The first constraint creates sentence structure. It is a partial order of words, that is (roughly) an ordering in which some words are higher or lower on some scale than others, while some are neither higher nor lower than others. The partial order holds between word occurrences in utterances. It determines all sentences, but is overt only in a subset from which, however, all other sentences can be derived. Grammatical relations can be defined in terms of it.

Consider a few words, in very short sentences—for instance, the words that can occur with *sleep*. In short sentences there are, as noted above, certain words that will occur with *sleep: man, child,* even *tree, earth (sleeps under a blanket of snow)*, etc. Some others are very rare here, e.g., *stone* or *coat* (in *The coat sleeps in the closet)*. In such short sentences there are other words that simply do not appear before *sleep* except metalinguistically, in discussions of grammar: for instance *sleep sleeps* or *is sleeps* or *because sleeps*. So each appearance of *sleep* occurs with any one word of a certain (tentative) set, but not with other words. The words that are not excluded will be called, for reasons to be seen later, the argument set for *sleep*. The same argument set, with different frequencies for individual words, will be found before *fall* or *fell*, *walk*, and many other words (*children fell*, *stones fell*, but not *because fell*).

Next, the word *wear*. *Wear* will be found to occur with two words of this same argument set: *A man wears a coat*; perhaps semijocularly *A tree wears bark*; but not *sleep wears run* or *the cause wears go*. We now consider another word, *assert*. *Assert* occurs with one word of the argument class that we already know: *The man asserts . . .*, *The child asserts . . .*, even a tree may assert something, and a painting is claimed to do so. But *assert*, like *wear*, also has another word following it: a man asserts something. That something, however, is not *coat* or *man*; one does not say *The man asserts a tree*, but rather, e.g., *The man asserts that a tree sleeps*. More precisely, *The man asserts that a tree sleeps* has *man* and *sleep* as the two arguments of *assert*, and then *tree* as the argument of *sleep*. Thus the argument of *assert* is any one word of the *man*, *tree*, etc., set and any one word of the *sleep*, *wear*, *assert* set (note also *The man asserts that I assert that he wears a coat*).

Finally, consider *entail*. It occurs with any two words of the *sleep*, *wear*, *assert* set, but not with any words of the *man*, *tree*, *coat* set. Thus *The man's wearing a coat entails the man's having a coat*, where the argument of *entail* here is the pair *wear, have,* and the argument of *wear,* as also of *have,* is here the pair *man, coat.* We do not find *John entails the man* or *John entails the man's wearing a coat.* There are other words that occur with different argument combinations but never involve any set other than these two: the roughly "concrete" nouns such as *man, earth, universe;* and the verbs and adjectives such as *sleep, wear, entail.* The latter class is found also to contain prepositions and conjunctions, and relational nouns.

The partial-order relation then is as follows: for each word (*man, sleep, wear, assert, entail*) there are zero or more classes of words, called its arguments, such that the given word will not occur in a sentence unless one word—any word—of each of its argument classes is present, in a stated position next to it. For *wear,* the pair *man, coat* are in its argument classes, but not the pair *man, walk* or *walk, sleep.* If a word, e.g., *man* or *tree,* has zero argument we call it a zero-level word; if it has a nonzero argument we call it an operator on that argument. In *The man asserts that a tree sleeps,* the *assert* is the operator on the argument pair *man, sleep* (forming *man asserts sleep*), while *sleep* is the operator on its argument *tree.* This is a dependence, with each word in the sentence requiring particular other words, more exactly any one word of particular classes: *sleep* requires the classes of *man, tree.* This dependence is a partial order. First, it is transitive: if *assert* requires the class of *man, tree* and the class of *sleep,* and *sleep* in turn requires the class of *man, tree,* then *assert* also requires (i.e., occurs only in the presence of) the second occurrence of the *man, tree* class, in this case *tree.* Second, there are words between which the relation does not hold: *man* does not require the set of *coat,* nor does *coat* require the set of *man,* though they may co-occur in many sentences (as above).

In the case where a word, e.g., *wear, assert, entail,* requires two words (as coarguments), it requires an ordered pair of them. Thus under *wear* the words that are relatively frequent in the first position (*man, child, earth*) differ from those that are frequent in the second position (*coat, blanket of snow*). And under *assert* the class of (roughly) nouns occupies the first position and the class of (roughly) verbs the second: *the man asserts a tree sleeps,* and not vice versa. Thus in addition to the partial order of operators in respect to arguments, there is a linear order among the coarguments.

The partial order is a constraint on word combination: it says that in the argument position next to a given operator the frequency (or probability) of certain words—those not in the argument class for that operator—is zero. Each satisfaction of the partial order, i.e., each word sequence in which all the source words have their requirement satisfied, is a sentence. Hence when a higher operator has another operator as its argument, e.g., when *assert* has *sleep* as its argument (*assert that a tree sleeps*), then the higher operator is acting on a whole sentence (*a tree sleeps*), which is then a component sentence of the bigger sentence being created by the higher operator (*He asserts that . . .*). The partial-order relation has a meaning: as will be seen later, each operator is being said about its argument, so that the meaning of the partial order is roughly predication.

While this dependence is based on the word combinations in a particular body of data, it is intended to predict the combinations in any utterance of the language except insofar as transformations (the reductions in the third constraint, 1.6) alter the shapes (and apparent presence) and positions of words. (The linear positions will be discussed in 1.7.)

Now, this dependence relation has an important property. If we ask what determines for each word which word class it requires as argument, we find that the required words are identified by what they in turn require. In our sample word classes, *man, tree, earth*—mostly concrete nouns—can be defined as requiring null, i.e., nothing. Then *sleep, fall*—mostly what we might call "concrete" intransitives—require a word of that class, that is, they require a word that requires null. And *wear, find*—"concrete" transitives—require two words that in turn require null. But *assert* requires an ordered pair from different classes: one word that requires null and one that requires something (such as *sleep, entail,* or *assert* itself). And *entail* requires two words each of which must require something (such as *sleep, assert,* or *entail* itself). There are here three levels of requirement, which indeed are inescapable: There must be, in the language and in each sentence, at least one zero-level argument that requires nothing, for otherwise one couldn't have any words in a sentence. There must also be at least one first-level operator that requires only words that require null, for nothing else could enter a sentence after the zero-level words; this includes both *sleep* and *wear*. And there would have to be second-level operators, at least one of whose requirements is a first-level operator, if we are to have any sentences beyond the elementary ones; this includes both *assert* and *entail*.

Thus the relation that imposes the partial order is not just the dependence of a word on a stated class of words, but the dependence of a word on the dependence property of words. This is the kind of relation that can define a system without recourse to any externally defined elements; it has the property of a mathematical object. Having come to this, it is worth noting that the language elements involved, namely words, have indeed no inherent property that has to be used for sentence construction. The sounds of words are not related to their meaning or their combinability, and are even dispensed with in writing (especially in pictographic writing). Even the meanings of words, as will be noted in the third lecture, are in part determined from their combinations rather than purely from their identity. From the viewpoint of the partial order, then, the word occurrences form a set of arbitrary elements closed under the dependence-on-dependence relation, with every combination that satisfies this relation being a sentence.

The importance of this relation will be clearer when it is seen that almost all further classes and operations in language, and almost all language meanings, are formulated on the constructions resulting from this relation. It should be noted that the operator-argument relation produced by this dependence has important similarities to functors in categorial grammar in logic. The differences arise from the different purposes in a syntax of logic and a syntax of natural language.

1.5 The Likelihood Constraint

We have obtained the gross structure of sentences. It is still necessary to describe how a particular word is chosen for a sentence, how certain combinations are more likely than others. This is done by a second constraint, on word likelihood. Whereas the first constraint creates sentence structure, the second specifies word meanings. It does not necessarily create meaning, since many words with their meanings must have been in use singly before being used in sentences, but it specifies meaning to any detail desired, and it enables a word to extend its meaning, and to have different meanings in different operator-argument environments. The first constraint set probability = 0 for words outside the required class in argument position; this leaves room for any probability >0 for words of the required class. Nothing says that all words must have equal frequency in respect to their operator or argument, or that the frequency must be random, or must fluctuate. In fact we find in language that each word has a particular and roughly stable likelihood of occurring as argument, or operator, with a given other word, though there are many cases of uncertainty, disagreement among speakers, and change through time. These roughly stable likelihoods, and especially the selection frequency, which will be mentioned in a moment, conform to and fix the meanings of words, as will be seen in the third lecture.

We speak here of likelihood under an operator (or over an argument), in the sense of estimated frequency or probability per fixed number of occurrences of that operator (or argument); no one has actually counted the frequencies of various words in argument position under another word. Nevertheless it should be noted that counting such frequencies over a small sample of the language is not as impossibly vast a task as it might seem to be, and this because we are not speaking of frequency in respect to other words in arbitrary sentences but only in the word pairs or triples in operator-argument relation, which is the elementary sentential structure and the sentential component of all sentences, and which constitutes the great bulk of meaning-characterizing, roughly stable relative frequencies.

Each word has a somewhat fuzzy selection of other words that are more likely than average to occur in the position for its argument—that is, more likely than would be expected if the occurrences were random or equal in frequency. Under *sleep*, this holds for hundreds of words such as *man* and even *tree*, in contrast with *earth* rarely, *stone* or *universe* even more so. The set of words having this higher-than-average likelihood is called the selection, in this case under *sleep*. The central meaning of a word is given by (the meaning of) the selection of arguments under it or of operators over it.

In addition there exist words with exceptionally high likelihood, and this on several different grounds. A word may have high likelihood as a total of many ordinary likelihoods, if it is in the selection of exceptionally many operators. For example, in the case of the indefinite *something, someone,* virtually every operator in the language will readily accept one or the other of these as argument; hence their total frequency is high. Or the high likelihood may be relative to a particular operator. For example, word repetition in certain positions is especially frequent under certain conjunctions. Thus *Schnabel played Beethoven but nevertheless Schnabel composed modern music* is more likely to be said than *Schnabel played Beethoven but nevertheless Berg composed modern music.* For the second subject to be the same as the first (here, *Schnabel*) is more frequent than for it to be any other one word, e.g., *Berg.* Under *and* or *but* it is more likely that some word in a given position in the second sentence be the same as in the corresponding position of the first sentence than that all the words be different. We will see in the third constraint (1.6) that high-likelihood words may be reduced, even to zero; and indeed these repetitions in corresponding positions can be zeroed, yielding, e.g., *Schnabel played Beethoven but nevertheless composed modern music.*

Some high-likelihood word occurrences are recognized to be such by the very fact that they have been reduced. For example, under *expect* we find the second argument to be either a sentence, as in *I expect them to leave* or *I expect them to be here* or *I expect that John will be here,* or else a zero-level word (a noun) as in *I expect John.* By the theory, *expect* should have one dependence or the other, but not both. Could one of these therefore be just a derivation from the other? Indeed we can find a way of deriving the noun object (*John*) from the sentence object (*John to be here*). There is a small set of words, roughly synonymous under *expect,* that may have such high expectancy under *expect* as to be zeroable there: *arrive, be here, show up,* etc. To say that *I expect John* is reduced from *I expect John to be here,* or *to show up,* accords not only with the meaning but also with various grammatical facts. For instance, in *I expect John momentarily* neither the expecting nor John is momentary, but the showing up is. Also, there is a telling relation between the connection of the noun object (*John*) to *expect* on the one hand and to *show up* on the other. The grading of likelihoods for noun objects under *expect* (from *I expect John* to *I expect dinner* to a rare *I expect New York* to almost never *I expect time*) is the same as for noun subjects under *show up* (from *John will show up* to *Dinner will show up* to *New York will show up* to almost never *Time will show up*). Instead of repeating under *expect* the same graded likelihoods that we have under *show up,* we need merely say that the second argument of *expect* is never a noun but only a sentence, but that if the operator in the sentence is *show up* or *come* it is zeroable (when under *expect*). If we say that *I expect* X *to show up* is reduced to *I expect* X, then *I expect time* is hardly ever said because *Time will show up* is hardly ever said. All this shows that *be here, show up,* etc. have special likelihood or otherwise favored status in the second arguments of *expect,* and are zeroable there.

There are also words with exceptionally low likelihood in particular situations; an example will be given later.

This constraint, of different likelihoods for different arguments, restricts the equiprobability of words; it specifies that for words with probability > 0 in the argument position of a given operator, some have higher than average or very high frequency and some close to zero.

1.6 The Reduction Constraint

The third constraint makes existing sentences more compact. It consists, for each language, of a few specifiable types of reduction, even to zero, in the phonemic shape of particular word occurrences. First, the domain of reduction: what is reducible is the high-likelihood (or otherwise favored) material. Certain words that have exceptionally high likelihood or special status in a given position are reducible; an example is zeroing the repeated corresponding words under *and;* and of *show up* under *expect.* The words that have highest likelihood, i.e., are expectable, in a given environment contribute little or no information when they enter there, in the information-theoretic sense (as will be noted in the third lecture). It is relevant that reduction takes place in several different high-likelihood and special status situations. This suggests that what determines reducibility is not simply high frequency but low information, which is the common property of all of these situations. Note that the ability of the hearer to supply the zeroed word shows that the to-be-zeroed occurrence of the word carried no further information that had to be given by the speaker.

This suggestion is supported by the fact that words that have exceptionally low likelihood in a particular operator environment can, when they do occur there, block reductions that would otherwise take place there. To see an example, consider that one can say *He has truly left* but not *He has falsely left,* and one says *He has certainly* (or undoubtedly or *possibly*) *left* but not *He has uncertainly* (or *doubtfully* or *impossibly*) *left.* It is not that *false, doubtful* cannot be said of *He has left*: we have *That he has left is false,* or *doubtful,* as well as *true.* But it can be shown that *He has truly left* is a reduction of (roughly) *He has left; that he has left is true,* where the second sentence has the semicolon intonation of a subsidiary sentence. Then *He has falsely left* had to be reduced from *He has left; that he has left is false.* Now it is reasonably likely for a subsidiary sentence to modify or weaken the primary sentence, especially with something like *at least,* as in *He has left; at least his having left is possible,* which is then regularly reducible to *He has possibly left.* But it would be rare for the subsidiary sentence to seem to contradict the primary, as in *He has left; his having left is uncertain* or *false* (where also no *at least* occurs). (Note that one can nevertheless say *He has, improbably enough, left,* where the source is somewhat different, though even here one cannot say *He has, falsely enough, left.*) It is clear that we are dealing here not with the basic sentence-making partial order but with details of likelihood for different words within the source sentences. Where reduction is not carried out, in the (1) *He has left; his having left is false* case, that seems due to the low likelihood of having the second argument under the semicolon conjunction deny the first argument (*He has left*). One can say the full form (1), but one does not reduce it. Reducibility, then, can be blocked by exceptionally low likelihood.

After seeing what gets reduced, we can note briefly the physical content of reduction. First, certain words of broad combinability are reduced to affixes: e.g., the (mostly) adverbial suffix *-ly* was reduced from an Old English *lice*, "in body, in form" (so that, e.g., *quickly* is *in quick form*). Second, repeated material can be reduced to pronouns or zero, as under *and*. Third, words of total high frequency (i.e., very broad combinability) may be zeroable, as in the case of *something, someone, one: To err is human* is from *For one to err is human* (which regularizes the infinitive clause into being a whole sentence form). Lastly, words of highest expectancy relative to a particular operator are zeroable under it, as in *show up, arrive* under *expect*.

Each reduction can be shown to take place as soon in the making of a sentence as the conditions for it are satisfied, before any further operator acts upon the affected words. Thus reductions are ordered in the partial order of word entry in sentence making. The reductions do not alter the presence of a word, only its shape and visibility. This is seen in the fact that the modifiers of a zeroed word can remain in the reduced sentence and make sense in it: e.g., in *I expect John momentarily*, the *momentarily* makes sense only as modifying the no-longer-visible *show up*. A reduction does not change the partial order of the affected words in the sentence, and changes little or nothing of their likelihood relations to the other words. What this reductional constraint does do is to restrict certain sequences of words, or rather of phonemes: it says that given the word combinations created by the first two constraints, certain words appear as different phoneme sequences or don't appear at all. One might argue that this constitutes a change in previous constraints rather than an addition to them. But many reductions (in English almost all) are optional, so that the source sentences also exist as possible sentences (even if some are not normally said) and must be provided for in the first two constraints. Hence the reductions are most efficiently accounted for as further constraints deriving the reduced form from the source, rather than as an independent description of the word combinations of reduced sentences. Even if many reductions were not optional, so that the source sentences were lacking in the actual language, it might well be a better organized and more efficient description first to obtain the principled source forms and then to add the effects of reductions.

To see that the reduction applies not to a word as such but to a word occurrence in a high-likelihood position, note that in colloquial English, where *going to* can be reduced to *gonna,* we can find *I'm gonna make it* from *I am going to make it,* but not *I'm gonna the next room.* The reason is that before nouns, *going to* is at selectional frequency only before certain ones of them (*New York, the next room,* but not before *word* or *time*); but before operators, *going to* is at selectional frequency before all of them (*go, make it, speak up,* etc.). Hence *going to* has total high frequency only before operators, and is reducible only there.

1.7 Linearization

One further step has to be considered before the three constraints that we have seen can produce the sentences of a language as they are said. Since the relation that makes sentences out of words is a partial order, while speech is linear, a linear projection is involved from the start. Every language has one or more normal linear projections. In English the operator is said after the first argument, so that *wear,* which has two arguments, appears between them as in *Men wear coats.* When there are complex combinations of operators upon operators this linearization can lead to ambiguities, as also in the usual mathematical notation. Such ambiguities can be avoided by parentheses, which speech does not have, or by a Polish notation in which the operator is either before or after its arguments but not between them.

The fact that the linear order is a separate step in sentence making, a projection of the original order, leaves room for alternative linear orders to exist without prejudicing the grammatical relations, which are created by the partial order. We will see that this yields the grammatical source of modifiers, which are very important in grammar.

There are two main alternative linearizations in many languages, including English. One of them is "fronting": bringing a word toward the front of the sentence, especially when it is, so to speak, the topic of the sentence. An example is *John I have long distrusted* from *I have long distrusted John.* The other is to interrupt a sentence with a whole subsidiary sentence. For instance, in *The opposition was unprepared* an interruption could yield *The opposition—so John said—was unprepared.* One can interrupt a sentence at various points. Especially important is a case that uses both fronting and interruption. It arises when the interrupting sentence begins (possibly as a result of fronting) with the same word as that before the interruption. For instance in *I believe that John is responsible* one can insert *John I have long distrusted,* yielding *I believe that John—John I have long distrusted—is responsible.* In this situation the adjacency of the repeated word, the two occurrences of *John,* enables the repetition to be reduced to a pronoun—*who, whom, which,* etc.—yielding *I believe that John, whom I have long distrusted, is responsible.* This is not a simple matter because language does not have an internal counting system; it does not have an addressing system for words in its sentences, and can therefore express cross-reference only under favorable conditions. When two occurrences of ostensibly the same word are next to each other, one can easily identify them, to say that the two are the same. If the two words occur at a distance, it is very hard to say in a general way what is the distance between them, and so to identify the antecedent of the repetition. So it is particularly in this situation, when one has reached *I believe that John—John I have long distrusted— . . . ,* that one can reduce the second occurrence of *John* to a "relative" pronoun *whom.* This is the birth of the relative clause, from which all modifiers in English are derivable. It is a very important construction, and in the present analysis it arises from linearization being a separate step over and above the partial ordering, which leaves room for certain alternative linearizations.

It remains to consider the status of linearization among the constraints. If the partial order is taken to be just the relation of arguments to their operators, then the linear order among coarguments is an additional constraint (with its own "grammatical" meaning), and the linearization of the partial order (which contributes no meaning) is an additional constraint only in that it chooses one out of a number of possible linear projections of the partial order. If we think of a standardized measuring of departures from equiprobability, then the linear form of the partial order has to be referred to all the time, for the changing linear transitional probabilities of phonemes or words in utterances. The alternative linearizations, which contribute a grammatical meaning, can be considered an additional constraint, since the most reasonable and efficient description of them may be as departures from the normal linearization, i.e., as additional steps over and above reaching the normal linearization.

In terms of the process of sentence making by these constraints, we have the following: the likelihoods and the reductions and the main alternative linearizations are defined on the operator-argument relation, i.e., the partial order; and languages do not in general have markers or an addressing system sufficient for any convenient carrying out of these relations on the linear projection of the partial order. We may therefore have to think of likelihoods and reductions being carried out before linearization. There is little or no evidence that the alternative linearizations are carried out as added permutations, after the normal linearization is in place; these should be thought of as alternatives in the linearization step.

1.8 Properties of the Base

We have seen the constraints on word combination: the partial order of word dependence that created sentence structure, the likelihood inequalities that fit word meanings, the reduction of high-likelihood word occurrences, and finally the linearizations. Each acts on the resultants of its predecessor. The constraints partition the set of sentences into two major sets. Without reduction, they create a base set from which all other sentences are derived. What is important here is that neither the base set nor the other set, the derived (reduced) set, is merely a residue of the other. On one hand, the structure of the base set is not just a description of all those sentences that could not be derived from something. On the other hand, the derivations are not just any change needed to obtain the remaining sentences from the base set. Rather, the base set and the reductions each have simple and understandable structures on their own terms, and it is a nontrivial result that the whole set of sentences is characterized by just these two structures.

Now we will look at a few properties of the structures as they result from the constraints, first of the base, and then of the set of the reduced sentences. In the base virtually all sentence words are simple, not composite, since affixes are generally reductions of next words; this makes the formulation of the base much easier. Few words have more than one defining dependence. If we have a word like *expect* with two types of objects as in *I expect John to come* (or *to go away*) and *I expect John*—one object a sentence, the other a noun—we usually find non–ad hoc ways of deriving one from the other. In each language the dependence defines just a few large word classes, a bit more than in our examples above: ones requiring null (zero-level, mostly concrete nouns), ones requiring one word requiring null, or requiring two such words (these are intransitive, and transitive, first-level operators). (One can say that *put* requires three words requiring null: *I put the book on the table,* not *I put the book.*) Then there are words that require one word that requires something non-null, i.e., requires one operator as its argument (e.g., *probable*), words requiring two such (e.g., *entail, because),* words requiring one zero-level word and one operator (e.g., *assert*), and some others; all these are second-level operators. These are the only kind of word classes defined in the base. The partial-order composition of each base sentence is transparent in the linear form. And, as will be seen, the base suffices for all the information carried by the language.

1.9 Properties of the Reduced Sentences

The sentences outside the base have their own structure. As was seen, they result from a fixed set of reduction types, and only when specified low-information, high-likelihood conditions are satisfied. Most reductions are optional, not obligatory; one can say both the long form and the short form. In English very few are obligatory. For those that are obligatory, the source sentences have dropped out. However, there is a way (in some cases historically evidenced) of finding for each obligatorily reduced sentence a suppletive source sentence, equivalent to it in meaning, which would take the place of the missing source sentence. An example would be to take *I am obliged for me to go* (replacing the nonextant *I must for me to go*) as suppletive source for *I must go.* Then every reduced sentence has at least one base sentence to which it corresponds; if the reduced sentence is *n*-fold ambiguous it has *n* corresponding base sentences. Thus, the base sentences exist, even if unwieldly in form and so never used (so long as they do not violate the vocabulary and grammar of the language).

Reduced sentences do not differ in meaning from the sentences from which they are derived. Reductions do not even eliminate the presence of the reduced word; they only change its shape, as in *going to* reduced to *gonna,* or they reduce the phonemes to zero. Hence the meanings of all the reduced sentences are available in the base sentences. This means that the simple structures of the base carry all the information expressed in the language, so that the notorious complexity of grammar, most of which is created by the reductions, is not due to complexity in the information and is not needed for information.

The set of reduced sentences has structural features

lacking in the base. There are marginal sentences due to fuzzy domains of various reductions. There are ambiguous sentences in the reduced set, made by degeneracy in reducing: different reductions from different source sentences may yield the same overt word sequence. But that word sequence preserves the two meanings of the two different sources; that is what makes it ambiguous. The reductions also create many word subsets when the domain of a reduction is not the whole of the base word class but only a high-likelihood section of it. In many languages the reductions have created morphology, as a reduced phonetic distance between certain words and their arguments, merging them into one composite word: *Childhood* is historically from *child-had* (from *had,* meaning "state"), from "the state of being a child." Reductions also create many special constructions. In the new phonemic shapes that the reductions make there are cases of similar grammatical relations having similar forms, some of which lead to structural patterns and—especially when the reductions are obligatory—to grammatical paradigms such as conjugations and personal endings on verbs. Such grammatical paradigms give prominence to particular grammatical meanings such as tense, plurality, and person; but all this is done in the reductions and all this has equivalents, from which it can be derived, in the base.

These are the major specific properties of the base and of the derived sentences, as they result from the constraints. They give language its form and its semantic capabilities. The more general properties of language, including its general mathematical structure, will be noted in the last lecture.

1.10 Methodological Summary

Before leaving this subject a methodological summary may be in order. What was presented here was not the analysis of a language, but a theory. But rather than give arguments for the theory, what was done was to show how it satisfied a distinguished "base" subset of sentences and how it built the remaining sentences out of the base. The initial sentences of the base did not have to be justified with the criterion of grammaticality—which is a circular criterion in any case—because its sentences are simple and well-established ones which were not selected for the purpose of the theory, but were distilled out of the other sentences by a formal simplificatory procedure (reversing the presumed reductions), by virtue of which the other sentences could then be derived from the base. This procedure, of recognizing and undoing intersentence transformations or reductions, has the necessary property that the source to which it leads is never more restricted than the given sentence, so that we recognize the base by its containing the least-restricted sentences of the language.

In any case, an essential problem about language is what differentiates the word combinations in it from those that are not in it. This means that the orderly investigation of constraints, whether as presented here or equivalently, is part of any language analysis. The fact that a great deal of structure, some of it mentioned here, comes as a corollary of these constraints indicates that the constraints are central to language. This centrality is indicated also by the fact that each constraint that contributes to the structure of a sentence also makes a fixed contribution to the meaning of that sentence. And it is also indicated by the fact that the first constraint creates a mathematical object—not of abstract concepts, but of the actual sentential word occurrences—as the fundamental structure of sentences.

It should be noted that while the picture of language presented here may seem too reductive, in that intricate structures are defined in terms of relatively simple constraints, this is not a reductionist view of a system as being nothing more than its components. Sentencehood is not just word choice, but a new relation—dependence on dependence—on words; and the individual sentence exemplifies not just sentencehood but a likelihood relation on the individual dependent words.

2

Science Sublanguages

2.1 Subsets of Sentences

HERE ARE introduced several related subjects: subsets of the set of sentences, sublanguage of a language, the metalanguage of a language, and finally science languages, consisting of formulas written in symbols—all as they relate to language structure on the one hand, and to the information of science on the other. This family of subjects is entered by noting that in the syntactic theory presented in the first lecture the very first step that is taken toward sentence making already creates a full sentence structure. No complex and graduated construction is needed for sentencehood: the simplest application of the dependence relation (a first-level operator on zero-level arguments) makes complete (elementary) sentences, such as *John wins*. Furthermore, these are elementary in respect to the other sentences not merely in that the other sentences can be derived from them, but also in that the other sentences actually contain them as components. Once the elementary sentences are formed, all further sentences are obtained by second-level operators on the first-level ones; but there is also a relation between the further sentence and the elementary sentences it necessarily contains. In this way, after the first step of elementary sentence formation, the grammar can be stated in terms of relations—of inclusion and expansion—among sentences.

In respect to this, other relations among sentences are introduced, in particular the partition into elementary sentences (*John is here*), other unreduced sentences (*I expect John to be here*), and reduced sentences (*I expect John*). Each subset of sentences has certain similarities among its members, and certain relations to other subsets. A sentence can thus be characterized not only by its composition, but also by its subset membership.

A subset of the sentences of a language constitutes a sublanguage of that language if it is closed under some operations of the language: e.g., if when two membes of a subset are operated on, as by *and* or *because*, the resultant is also a member of that subset. The elementary sentences do not constitute a sublanguage, because any operation on an elementary sentence makes it no longer elementary. The base (i.e., unreduced) sentences, elementary and not, constitute a sublanguage, and so do the reduced sentences; the latter set has the property of an ideal, since combining a reduced sentence and an unreduced one yields a sentence that has to be considered reduced. The structure of each sublanguage, stated in its grammar, is appropriately different from that of the whole language.

Another important structurally distinguished sublanguage is the metalanguage. Since words can refer to words no less than to other things, we can investigate all those sentences of a language that refer to words of the same language. The set of all such sentences in a language is the metalanguage of that language: it identifies the elements of the language and their combinations and relations, and the meanings of all these. Within this metalanguage, a compact system of sentences sufficient to describe all sentences of the original language is a grammar of that language. Now, it has been noted that natural language has no external metalanguage capable of describing it. But a subset of its own sentences constitutes a metalanguage, and a grammar, of it.

This metalanguage, to be called here M_1, initiates an interesting regress of metalanguages. First, M_1 has a structure different from the whole language. The zero-level arguments in M_1 are the sounds, words, and constructions (word combinations, sentences) of the whole language; its first-level operators are *is a word in the (whole) language, is a sentence in it, is phonemically distinct from, is a variant of, occurs in, is next to, is reduced,* and the like. Its second-level operators include *is more frequent than* and the like. The description of the structure of M_1, which is its grammar, is given in the metalanguage of M_1, which will be called here M_2. M_2 is not contained in M_1; it is a separate set of sentences from some whole language, for example from the language that M_1 was describing. M_2 itself has a structure, somewhat different in its syntax from the structure of M_1. The zero-level arguments of M_2 are the words, word classes and constructions of M_1, e.g., *phoneme, indefinite noun, elementary intransitive operator, probability grading.* The first-level operators of M_2 do not include, e.g., *is a variant of, is reduced* (because the metalanguages may avoid reductions in their own sentences), and the second-level operators of M_2 will not include *is more frequent than,* because the frequency of the combinations in the sentences of the metalanguage (as a science language) is not important, as it usually is not in languages of science. This grammar of M_2 is in M_3, which is the metalanguage of M_2. The structure of M_3 is slightly different from the structure of M_2. The difference is primarily in its zero-level arguments. These arguments in M_3 are all the terms and sentences of M_2, i.e., all those needed to describe the structure of M_1. They include *word, relation, classification, and order.* However, if we take this grammar of M_3 as being a case of M_4 and study the structure of M_4, we find that it is identical with the structure of M_3. The referents are different: the elementary arguments of M_4 refer to the expressions in M_3, while the elementary arguments of M_3 refer to the expressions of M_2. But the actual words and sentences used in M_4 can be the same as those used in M_3. Thus in their referents there is an infinite regress of metalanguages, as expected, but not in their actual form.

In effect, M_1 is a science language, differing from the whole language; M_2 is a grammar of a science language, itself a science language but differing from the one it describes; M_3 is a grammar of a grammar of the first science language, still differing; but M_4 and its successors no longer differ in form. This case is offered as an example of how results can be obtained from investigating the structure of sublanguages.

2.2 Subject-Matter Sublanguages

The base, reduction, and metalinguistic sublanguages are all defined in respect to language structure. In addition, the metalanguage is also characterized by its subject matter. Many other subject matters support distinguishable sublanguages. Reports and discussions in well-structured aspects of the world show limitations on word use that are sharp enough to constitute constraints on word occurrence; this is especially evident when what is said is limited to what seems relevant to the field. When such material is investigated by means of the dependence criterion of the partial order, we obtain for each such subject matter not the concrete nouns, intransitive and transitive operators, etc., of the whole-language grammar, but different word classes, different sentential constraints, and other different grammatical properties. The same methods yield two kinds of grammar when they are applied to two different data conditions. It would not have been possible for the same methods that yielded the whole-language grammar to yield a different grammar for sublanguages if we had built the whole-language grammar out of fixed classes such as nouns, verbs, and adjectives. This is so because the classes that turn out to be relevant for the subject matter are not noun, verb, etc., but subsets of these, and even more because some subject-matter elements or classes cut across the whole-grammar word classification (as will be seen below, and in the internal conjunction in the immunology sublanguage).

When we make a separate grammar for a given subject matter, we find not a general dependence on dependence, but specific sets of arguments occurring only under particular sets of operators. In the whole of English *Polypeptides were washed in HCl* and *HCl was washed in polypeptides* both sound like grammatical sentences, even if we know that the second would describe a rather peculiar activity. But in laboratory reports the second would simply not occur. Thus we would not have noun-verb-noun sentences and the like, but instead a set of operators like *is washed in* would have as its ordered pair of arguments a set of words for (roughly) molecules and cells and tissue and a set of words for acids, water, and the like. Other operator subsets would have other argument subsets.

The metalanguage of each such sublanguage, its grammar, would not be a subset of it as in the case of the whole language. Rather it would be just some other set of English sentences, external to the sublanguage it describes. Having an external metalanguage makes a difference. It means that defining the word sets and structure of the sublanguage is not restricted to internal combinatorial regularity in the way we have seen for the whole language. Any set of words or word sequences can be defined for the sublanguage in the metalanguage of that sublanguage. In particular, whole phrases can be listed as members of a word class, i.e., as indivisible elements. For example, *appears in* and *is found in* may both be taken as operators on the argument pair *molecule, cell,* as in *Antibody appears in plasma cells* and *Antibody is found in plasma cells;* but for the whole language *is found in* would have to be analyzed as a passive going back to *Someone finds antibody in plasma cells.* In a sublanguage, some of the members of a class are pure synonyms (something that is rare in the whole language), arising from the fact that more words are available in the whole language than are needed in the sublanguage: e.g., the *appears* and *is found* just mentioned are synonyms in the sublanguage presented below.

Though the sentences of the sublanguage are a subset of the sentences of, say, English, the grammar of the sub-language is not a subgrammar of English. The sub-language has important constraints which are not in the language: the particular word subclasses, and the particular sentence types made by these. And the language has important constraints which are not followed in the sublanguage. Of course, since the sentences of the sub-language are also sentences of the language, they cannot violate the constraints of the language, but they can avoid the conditions that require those constraints. Such are the likelihood differences among arguments in respect to operators; these likelihoods may be largely or totally disregarded in sublanguages. Such also is the internal structure of phrases, which is irrelevant to their membership in a particular word class of a sublanguage.

Aside from the purely structural characteristics of these sublanguages, there is an important property in respect to information. It will be noted in the third lecture that when the word combinations of a language are described most efficiently, we obtain a strong correlation between differences in structure and differences in information. This correlation is stronger yet in sublanguages. The more stringent boundaries and tighter interrelations in the subject matter are reflected in sharper correspondence between word combination and information. Indeed, a major interest in analyzing the language of science is not so much that such formal or quasi-formal systems exist, as that they can be used to characterize the information in the given sciences. For this purpose it is all the more important that the structure be worked out with no assist from our view of its information, since we have here an opportunity to reach an objective and independently obtained structuring of the information. This means a purely word-combinatorial investigation. Nevertheless, since the material covered in a science sublanguage is far narrower than what is available for a whole language, limited shortcuts and educated guesses have to be used in practice, as long as they do not beg the question of information. The test of the grammar comes when the description made from a given amount of material remains adequate as the amount of material increases.

2.3 Science Sublanguages

When the subject matter in question is a subfield of science, we obtain a complex linguistic system that can present precisely the information of the science, losing nothing from the original information as given in the whole language. In the science sublanguage we find distinguished ways of indicating the structure of the information, and the disagreements in information, and changes in it.

Let us consider first a specific case. I will sketch briefly an analysis of representative research papers (selected by T. N. Harris and S. Harris) in immunology, which was carried out by Michael Gottfried, Thomas Ryckman, and myself, and more recently also with Paul Mattick, Jr.; the analysis was tested for French by Anne Daladier.[1] The period covered was c. 1935–1966, when the field was far smaller and more inspectable than it is now, and when it had a central research problem of determining which cell was the producer of antibodies. There was also a controversy about whether it was the lymphocyte or the plasma cell, both of the lymphatic system. After it was shown, by electron microscopy and other methods, that both cell types produced antibodies, the controversy was resolved by the understanding that the two cell names indicated different stages of development of the same cell line. The purpose of the analysis was to see if one could represent, by formal procedures, all the information in an orderly and usable way, if one could locate in the sentence structures (and could characterize structurally) the changes in information over the years, and if one could locate and characterize the disagreements.

[1]Z. Harris, M. Gottfried, T. Ryckman, P. Mattick, A. Daladier, T. N. Harris and S. Harris, *The Form of Information in Science: Analysis of an Immunology Sublanguage*, Boston Studies in Philosophy of Science (Dordrecht: Reidel, 1987).

In its barest outline, the sublanguage consisted of the following: By listing how words occurred with each other in sentences of the articles, and collecting words with similar combinability into classes, some fifteen classes were found, chiefly for *antigen* (G), *antibody* (A), *inject* (J), *tissue* (T), *cell* (C), then for verbs (operators) between A and C (e.g., *appear in, produced by, secreted by)*, between G and T (*move to*), between C and C (*similar to, develops into*), and verbs on T or C (T *is inflamed,* C *proliferates*). These appeared in fewer than ten major sentence types, chiefly those exemplified by *Antigen is injected into body, Antigen moves to tissue, Cells or tissue change* or *have some property, Antibody appears in cell* and *Cell is the same as,* or *develops into, another cell.* For reasons that will be apparent later, each class is written with a capital letter, so *Antibody appears in lymphocytes* is AVC.

The many sets of synonymous words, especially verbs, are considered to be just variant forms of a single word, and the variants are not indicated. The non-synonymous words within a class are marked by subscripts, as in *appears in* (and synonymously *present in, contained in*) V_i, *produced by* V_p, *secreted by* V_s. There are modifiers: on certain verbs, e.g., *not, increase, from . . . to, begin to, have a role in;* on certain nouns, e.g., *much, immature, family of* (cells). These are marked by superscripts on the letter.

Here are the findings in a bit more detail:

- The words for *antigen,* marked G, were either synonyms of it or names of various antigens.
- Words for *antibody* (A) were local synonyms such as

gamma globulin, indicators of antibody presence such as *agglutinin,* or names of specific antibodies. It should be noted that most of these synonyms would not be synonymous with *antibody* in other research—*gamma globulin* is a broader term than *antibody*—but locally, in these texts and word environments, there was no difference, and indeed in most cases what is meant, for example by *gamma globulin* here, is just *antibody.*

- There is a class of bodies, body parts, and animal names (B): *rabbit, footpad,* etc.
- *Inject* (J) is used between G and B; there are also local synonyms for *inject.* In this material *immunize* was used synonymously with *inject,* and *normal* was used to mean *not injected.*
- There is a large class of *tissue* words (T) with many different tissues and organs which have to be distinguished by subscripts: T_l *lymph,* T_n *lymphnodes,* T_s *spleen,* and so on. They all have certain common environments, with differences among them in respect to detailed environments.
- There is a smaller class of *cell* names (C): C_y for *lymphocyte,* C_z for *plasma cell,* C_b for *blast cells* and quite a few other names, some proposed by the investigator just for particular properties that were found.
- There is a class (V) of the verbs that occur between A and C (or A and T): V_i for *in* (synonymously *found in, present in, appear in; contain,* in the inverse order C–A) as in *Antibody appeared in the lymphocytes* (AV_i C_y), V_p for *produced in,* V_m for *pass through,* V_s for *secreted by* as in *Antibody is secreted by the cell.*
- There is a large class (W) of verbs and adjectives after T

or C, naming properties or changes of tissue or cell, with different subscripts: *tissue is inflamed, cell is mature, cells proliferate,* etc.

- There is a class (U) of verbs that occur between *antigen* and *tissue* or *cell: antigen moves to tissue, perishes in tissue.*
- Finally there is a class (Y) of verbs between *cell* and *cell: is similar to, is called;* there is also Y_c *changes into, differentiates into,* and some more special subsets such as Y_u *contaminated with.*

The major sentence types, with some indications of their examples:

- GJB, for *Antigen is injected into a body part* or *an animal.*
- $GU^{ft}TT$, for *Antigen moves from some tissue to some tissue* (the superscript indicates *from . . . to*).
- TW and CW, for *A tissue* (or *cell*) *has some property* or *undergoes a change.*
- AVC, for *Antibody appears in, is produced by,* or *is secreted from a cell.*
- CYC for *Some cell is similar to,* or *is called, some cell.*
- CY_cC for *A cell develops into another cell.*

In "donor" research, in which antigen is injected into one animal, whereafter lymphocytes are injected (transferred) from that animal to another, with antibodies then being sought in the second animal, an additional sentence type is found:

- $CI^{ft}BB$, for *Cells are injected from an animal into another animal.*

There is a special conjunction, internal to a particular sequence of sentence types, which is seen or is implicit in almost all occurrences of the pairs GJB and AVC. This is *thereafter* and its synonyms, marked by a colon. It often carries a time modifier, e.g., *three days, five hours*:

- GJB:^{t}AVC, for *Antigen is injected into a body part; three days* (or the like) *later antibody appears in cells.* Also *Antibody appeared three days after injection of antigen,* and the like. This conjunction takes different grammatical forms (e.g., *to* in *The cell contained antibody to the antigen*); all of these forms synonymously connect AVC (or CW or TW) to GJB.

It is now possible to mention the main advances obtainable here from the codification of the sublanguage structure.

Metascience material, giving the scientist's relation to the information of the science, can be separated off. Mostly, this comes out immediately, as the highest operator in a sentence (e.g., *as was expected* or *A and B have shown that*); but there are also cases where metascience operators combine with operators of the science language proper, as in *Antibody is found in lymphocytes,* which could be analyzed as reductive consolidation either from (roughly) *It was found that antibody is in lymphocytes* or from *We found antibody in lymphocytes.*

We obtain a gross framework for the information in the field: the class sequence formulas such as AVC.

We also obtain a representation for the specific information in each sentence: the individual formulas with subscripts for different class members and superscripts for modifiers, as in $AV^{r}{}_{p}C_{y}$ (*Lymphocytes have a role in the production of antibody*); the superscript r indicates participating in production as against actually producing.

We find, in this particular sublanguage, tightly knit sentence-sequences, marked by a colon (GJB:AVC for *Antigen injection is followed by antibody appearing in cells*), with possible insertions (GJB:GUT:AVC for *Antigen injection is followed by antigen moving to a particular tissue whereafter antibody appears in cells*) and alternative paths (GJB:TW for *Antigen injection is followed by a particular tissue being altered*).

We see how related research lines differ: In the donor research mentioned above, we have GJB_1: $CI^{ft}B_1B_2$:$AVCB_2$ for *Antigen is injected in animal 1; thereafter lymphocytes are injected from animal 1 to animal 2; thereafter antibody appears in lymphocytes in animal* 2 (subscripts here to distinguish the two animals).

Within most papers we find differences in sentence types between the Procedures, Results, and Discussion sections into which laboratory reports are usually segmented. The differences and what they may yield will be noted below.

We can locate change over time: First AVT is replaced by AVC. In the earliest paper (1935) we find T (and only at the end, briefly, C), because cell types were not then readily distinguished in the tissues. Later, a new sentence type, CYC, enters, when more cell types are distinguished and their similarities noted, and when the proliferation of cell names is controlled by saying that some different names identify the same cell.

We can locate unclarity, as when the proliferation of cell names is not supported by different properties (in the W class) reported for the differently named cells, with the unclarity being finally recognized by CYC sentences stating that these are names for the same cell.

We locate the disagreements, and see their structural status. These come out as symbol differences at specific points. The chief case here is that papers of one set have AV_pC_y (*Antibody is produced by lymphocytes* or *Lymphocytes produce antibody*), while papers of another set have AV_pC_z (*Antibody is produced by plasma cells*) and $AV_p^r C_y$ (*Lymphocytes have only a role in antibody production*) and even $AV_p^{\sim}C_y$ (*Lymphocytes do not produce antibody*), but do not have AV_pC_y (see above). The contradiction between $AV_p C_y$ and $AV_p^{\sim}C_y$ is overt.

We locate the resolution of the disagreement when $C_yY_cC_z$ (*Lymphocytes develop into plasma cells*) appears in the final papers. Sentences of the form CY_cC, stating that one cell was a later stage of the previous cell, were becoming frequent in the later papers as many cell names and cell stage names appeared in the course of various experiments. But the two contenders for antibody production, namely C_y and C_z, had never appeared as the argument pair for Y_c; that is, the development was not recognized as reaching from one antibody-producing cell to the other. When both cells were shown to be producing antibody the explanation (of their being in the same cell line) was expressed by extending Y_c to the C_y, C_z pair: $C_yY_cC_z$.

Such observations, which are made more available by

the inspectability of the formula structure, make room for useful interpretations. They also suggest questions; for example, in the case of the colon-marked sentence sequences, whether the occasional insert GUT (*Antigen reaches tissue*) is implicit in all the GJB:AVC in which it does not appear; also what is the precise relation, in respect to GJB, between the two alternative followers of it in GJB:AVC and GJB:TW (or CW). Finally, they suggest the possibility of more orderly and wide-ranging critiques, based on the formulas, in respect to change, unclarity, and disagreement.

2.4 Science Languages

As this point we can see what is accomplished by using symbols instead of the original English, French, or other words. It is not just a matter of convenience or clarity. The symbols enable us to avoid synonyms (each symbol can represent a set of synonyms), and also to disregard the internal composition of a class member if it happens to be a whole word phrase (e.g., *is found in* as an inverse of *contains* in V_i, or *left footpad* as a member of B). The subscripts give a standard form to the subclass status: e.g., in noun classes, for various specific antigens marked by subscripts on G; but also, in operator classes, for the various relations between a fixed pair of noun classes (thus relations between A and C include V_i for *appear in* and its synonyms, V_p for *produced by* and its synonyms, and V_s for *secreted by*, etc.). As for the superscripts, they make it possible to locate the modifiers on their sublanguage hosts, as they usually are placed in language, even in those cases in which they are not said in that position in normal speech. Finally, it is possible to omit grammatical requirements of the whole language that are irrelevant to the particular science, e.g., in some cases tense, or plural, or reductions, and more generally distinctions such as between left and right modifiers in language.

With this, we move from science sublanguages to science languages. These are systems of information formulas composed of symbols. They can be obtained in a precise way from the sentences in the original language, but they do not have to represent any grammatical features of that language that are irrelevant to the given science. Indeed, the analysis of French immunology papers in the research area discussed above reached the same formulaic representation as for the English papers. Science languages constitute a new kind of system. The main feature that a science language has in common with natural language is the partial order that creates predication and sentences, but this mathematics also has. Science languages also retain from natural language the relative clause construction, which yields modifiers (adverbs, adjectives); mathematical notation lacks that construction in general. Approximately like mathematics, science languages do not have likelihood gradings of words in respect to their particular operators; but they do have, instead, argument subclasses in respect to operator subclasses, which mathematics does not. And perhaps even more than mathematics, science languages avoid the reductions and resulting paraphrases characteristic of language; for example, both *Antibody appears in lymphocytes* and *Lymphocytes contain antibody* would appear as the same formula AV_iC_y. Again like mathematics, science languages differ from whole languages in having an external metalanguage (a subset of the whole language).

Most of the features are clear from what has been seen of the immunology language, but the absence of likelihood grading requires additional comment. It has been noted that in language as a whole almost every word has a unique subset of operators on it (or arguments under it) in respect to which it has "selectional" (higher than average) frequency. Over its whole range of operators (or arguments) this selectional subset characterizes the central meaning of the word. In respect to each operator many different arguments have different likelihoods. In science languages this does not seem to be the case in respect to first-level operators. One antigen indeed differs from another, but not, in most cases, in respect to being injected, moving to tissues, or involving lymphocyte response. The difference in meaning of the different antigen names is thus not related to difference in the immediate (first-level) operators to which they have selectional frequency (other than in definitional and other special sentences). Rather, it is related to farther operators, in expanded or conjoined or sequential sentences with which they occur.

In the language as a whole the frequency of a word that depends on words of a given dependence status (i.e., on words depending on null, or on words depending on something) is stated in respect to the whole class of words having that dependence status: thus *sleep, fall,* and *rise,* which require a word that requires null, are considered to have positive probability, no matter how small, on every single null-requiring word—*man, tree, coat, universe.* In contrast, in a science language, many words within a word's dependence-on-dependence class have probability = 0. Thus the arguments of Y_c (*develops into*) are two zero-level words (of the same capital letter class), i.e., two words that require null (e.g., *cell*), but not all such words. Many zero-level words, such as animal and body part names (B), have zero probability in argument position of Y_c (no *The liver develops into a spleen*); only cell names and occasional other words occur here as arguments of Y_c, and among them the differences of frequency in respect to Y_c do not seem to be structurally important. In this way, science languages have replaced the selection differences and likelihood grading over a whole dependence-on-dependence class by co-occurrence subclasses such as Y_c and C_z, depending on lists of argument words and no longer just on the dependence properties alone of the arguments; and within the subclasses, likelihood gradings seem to be irrelevant.

The science language is then a body of canonical formulas, representing the science statements after synonymy and the paraphrastic reductions have been undone. Its grammar states the class symbols (here, capitals), the class members (here, subscripts), the modifiers (superscripts), with the constraints on each, and with the combinations of them that constitute sentence types.

Within a science, one can state the structural and semantic connection of these formulas to any relevant formulas in other systems—mathematics, chemical compound notation, chemical reaction formulas—especially if the other formulas occur in the same texts as the language material, i.e., as the science language formulas. As was seen in the immunology analysis, the gross formulas show the framework of the science information, while the detailed ones represent the information carried by individual sentences.

The canonical formulas need not be just elementary sentences. In a given science some of the basic formulas may consist of particular higher operators on a particular kind of elementary sentence. Thus in investigating pharmacology texts it was found that a typical structure, exemplified by *Digitalis affects the beating of the heart,* consists of an elementary sentence *The heart beats* under an operator such as *affect, influence,* whose first argument is usually a drug name, possibly with a modifier that specifies a dose. This structure may indeed be an elementary sentence structure for pharmacology. A canonical formula of a science may also contain characteristically a pair or sequence of elementary sentence structures, as in the GJB: AVC of immunology.

The formulas of science languages have various other properties, aside from their gross structure. If metascience material is present, it is always the highest operator in the partial order of words in a sentence; this includes adverbial phrases such as *as was expected,* which are the highest operator of what was originally a secondary sentence on the given sentence. Some science languages have words and phrases for quantity, time, and spatial relations at stated points in the partial order: the immunology formula has time modifiers (superscript) on many occurrences of the internal colon conjunction, but not elsewhere.

There are also various relations of partial similarity and of sequential constraint among science language formulas. If we look at the typical division of experimental articles into Procedures, Results, and Discussion sections, we find that these are not just a matter of tradition, but are grammatically distinguishable. In the Procedures section, the immunology formulas identify particular antigens, the number and conditions of injection, the precise location of injection, what animal, and so on. The Results section has different sentence types, for example the GJB:AVC for *Antigen is injected, thereafter antibody appears in cell.* But there are certain words that are shared by the two sections, and what is more important than the mere sharing of words is the fact that the Results sentences are seen to be partially dependent on the Procedures sentences. Certain Results sentences could not appear if the Procedures sentences were different. This is hardly surprising, but the point is that it can be shown in the formula structure, and in a precise way.

A more important situation is found when we compare the Discussion section with the Results section. Whereas the Results section formulas are largely different from the Procedures ones, the Discussion section for the most part does not bring in new formula types. Rather, its sentences consist of Results formulas, often modified by using classifiers instead of specific words for objects and experimental conditions. These are, to a large extent, arranged in particular sequences and with particular conjunctions. This is faintly reminiscent of the syntactic conditions for proof in mathematics. It raises the question whether we could specify conditions on Results sentences, on what are the symbol relations that they may contain, and on their ordering and their conjunctions, such that particular arrangements of particular Results formulas would justify the assertability or reasonableness of the concluding formula on the grounds of the assertability or truth of the preceding ones. This is not a pious hope about reshaping science argument into mathematical proof. Mathematical proof depends on syntactical conditions that no science language can satisfy. But we have here the possibility of investigation into how Results sentences lead to justified conclusion sentences in the actual reports of the science, at least in those reports whose argumentation is considered satisfactory.

2.5 The Uses of Science Formulas

We will now consider the uses of these formulas, short of what further research may be able to do with them. We note first that since very many sentences, especially within the same section of an article, are cases of the same gross formula, the formulaic representation of a section of an article approaches a double array or even a tabulation. For example, in many articles the GJB:AVC formula is repeated many times, sometimes with occasional intrusion of some other formula. The progression of information through the section is achieved in the successive symbols in each position ("column") of the successive GJB:AVC rows. It appears in the different superscripts, which express the different modifiers, conditions, amounts, and so on, and in the different subscripts, which identify the different particular antigens, body parts, antibodies, and cells that are producing, secreting, etc.

The formula structure with its repetition forms the double array, which locates the information. We know where to look for the information, if it is there, and we know what form each item of information would have. In principle, the repeated formulas permit retrieval of the specific information, although making this structure into a computer capability requires a great deal of sophisticated work. In any case, the whole structuring as seen here makes the information inspectable even by the human eye. It locates omissions, as does all structuring and procedural representation, and it locates lack of information and imprecision—and we msut bear in mind that almost all research, other than mathematical, has imprecision at one point or another. It also makes possible precise comparisons of different documents in the science. And, to think of the future, it may become possible to analyze information in real time so that the tabulation and critique of past work can be used to affect ongoing work in the field, if the field does not change too fast.

Further, a canonical form for information may make it possible to specify and quantify differences between science languages or between the kinds of information carried in different sciences. We may be able to say explicitly wherein near subsciences differ. We may be able to see various relations among sciences; for example, which sciences are prior sciences of others. In general, a prior science would be one whose sentences are arguments in the sentences of the other science. As an example of this, consider in pharmacology how the physiology sentence type, as in *The heart beats,* was an argument in the pharmacology sentence type above, as in *Digitalis affects of the beating of the heart,* showing physiology to be a prior science to pharmacology.

We can see, in the different levels of the partial order of words in a sentence, the contribution of different sciences and different aspects of science. Each occupies different fixed places in the partial order of the words in the sentence types of a science. The metascience material, about the scientists' activity, is always at the top in this order within the sentence. The evidential status of the statement and the time/space/quantity relations and the like are at various intermediate levels. Then comes the specific event reported, the actions or relations of the objects of the science itself, and below that the objects of the science that are acted upon in these events. When present, the material from prior sciences is at the bottom.

In addition to all this, we may be able to see how and when a science advances and how it changes. Overall, science information is thus brought closer to being an exact system.

3

Information

3.1 How Sentences Carry Information

IN A consideration of language structure, the issue of information enters because language is clearly a carrier of meaning and information. In the present theory it enters even more directly, because as we approach a least grammar, with least redundancy in the description of the structure, the connection of that grammar with information becomes much stronger. Indeed, the step-by-step connection of information with structure is found to be so strong as to constitute a test of the relevance of any proposed structural analysis of language. This suggests that the components that go into the making of the structure are the components that go into the making of the information.

Meaning and information are not explicitly defined terms. We cannot investigate their structure independently and then compare or correlate that with language structure. But we can ask how words carry meaning and how sentences carry information. We will then consider examples of how an increment in structure yields an increment in informational capacity, i.e., in what information the structure can carry. It will then be possible to arrive at some general comments about the relation of form to content in language, and even about the structure of information as it appears in language.

First, we will look at how sentences carry information. It was proposed in the first lecture of this series that sentences are obtainable from a few constraints on word combination. Therefore, to see how sentences carry information, we ask how these constraints carry information. First of all, there are some constraints that have no direct relation to information: for example, the phonemic composition of words, or the spelling of words. There are some constraints that have a particular indirect relation to information, namely that they affect the access of the hearer to the information, either in respect to the time or ease of access. An example is in the reductions that change the phonemic shape, the sound, of words even to zero—though the words are still present, but in zero shape; the hearer then has to reconstruct the word, because its meaning is still playing a role in further word choice, even after the word has been zeroed.

There are also some constraints that express the speaker's attitude to the information that he is giving, again without changing the information itself. One such constraint is the fronting of words, bringing certain words to the front of the sentence in order to express the topic, the speaker's indication of what he is talking about. Another is bringing some sentences, as interruptions, into the inside of another sentence, which is a step in the process of making the second sentence into a modifier of a word in the first sentence: the substance of the information is the same but there is a change in the attitude to the information. There are also some constraints that create categories or relations of information without affecting the information itself. Such are some required reductions that create paradigms in grammar, e.g., the attachment of a plural morpheme to a noun if it is not singular.

In the whole system that has been found sufficient for grammar, only two of the constraints contribute to the substantive information of the sentence. These are the two constraints that create the base, i.e., the unreduced sentences in the language: the partial ordering of words which is the "predicational" relation of operator to argument, and the likelihood inequalities of words within the partial ordering (which distinguish word meanings). The contributions these constraints make to the meaning of their sentence is fixed for all the occurrences of these constraints.

There are other meanings expressed in speech—by intonation, by pauses, by the loudness and the rate of speech; but these do not combine in any regular way with the rest of language, and cannot be fitted into the structural system. There are also meanings expressed in irregular association with speech, but external to it, such as gesture. Any substantive (objective) information carried by all of these can be also expressed in language proper, but only via the above two constraints that create sentences.

3.2 How Words Carry Meaning

At what point do words get meaning? One should first note something that may not be immediately obvious, and that is that meanings do not suffice to identify words. They can give a property to words that are already identified, but they don't identify words. Another way of saying this is that, as everybody who has used Roget's *Thesaurus* knows, there is no usable classification and structure of meanings per se, such that we could assign the words of a given language to an a priori organization of meanings. Meanings over the whole scope of language cannot be arranged independently of the stock of words and their sentential relations. They can be set up independently only for kinship relations, for numbers, and for some other strictly organized parts of the perceived world.

In practice, the procedures that are effective for determining words are not semantic: that is to say, they do not rely on meaning. The direct procedure that establishes phonemes by distinguishing word repetition is based on the hearer's differentiation of speech sounds, and is not able for instance to distinguish homonyms (e.g, *heart* and *hart*) one from the other. The stochastic procedure mentioned in the first lecture, which exhibits the boundaries of words in an utterance, does not rely on the meanings, but only on the sequences of sounds, of phonemes or letters. In fact, correlating phoneme sequences with meanings is not enough to determine that something is a word. For instance, the phonemes *sl* have a certain hard-to-specify meaning initially in many words, as in *slide, slither, slick, slimy*, etc.; and *gl*, again initially, have a certain meaning in *glimmer, gleam, glow*, etc. But we cannot classify these sequences as morphemes (words or affixes), because their combinations with other morphemes are not sufficiently regular. There is even a phoneme sequence in English that used to be an affix—*le* at the end of *dazzle, nuzzle, juggle, frazzle*, etc.—but has spread irregularly into so many other words that it is hard to define it as a morpheme in English grammar today.

When the question of what is a word is left purely to semantic distinction, as is commonly done for homonyms within a word class, we find great difficulty in arriving at a principled decision. Thus, there are three ranges of meaning for the phoneme or letter sequence of *sound*: "noise," "a body of water," and "healthy"—the first two both nouns. All three are from unrelated etymological sources. These are considered different words. It may be a bit less obvious that the two nouns *shed* in *storage shed* and *watershed* (also of unrelated etymology) should be considered different words. The two verbs *will*, one meaning "in future" and the other "desire," are etymologically related, but are considered two different words, not so much for their minor difference in meaning as for their grammatical difference (the first having the properties of an auxiliary verb and a tense).

The problem of deciding whether a given phoneme sequence is one word with all its meanings, or two homonymous words, is made easier by the fact that words do not for the most part appear alone. They appear in particular environments—at least when they are at selectional frequency—and may have different meanings under particular operators or over particular arguments. The word *set* has a different meaning when it appears with *mind* or with *silverware* or with *theater.* The operator *divide* has virtually the same meaning as the operator *multiply* when its argument is a cell name: for a cell, to divide is to multiply. The dependence of meaning on the environing operators and arguments is related to the restrictedness of that environment: words that have no restriction on what operators they occur with (e.g., the indefinites) do not have different meanings under different operators. That is to say, having different meanings under different operators can be blocked by the grammatical property of unconstrained selection. In view of all this, homonymity could therefore be measured by the operator or argument range of a word as well as by how different its meanings are with the same operator or argument.

Furthermore, the operator and the argument range of a word changes through time, and with it the meanings of the word. Words extend into new combinations and drop out of some old combinations. It has been shown, most strongly by Henry Hoenigswald, that the extension of a word to new combinations is determined, not simply by the meaning of the word, but by the way the existing combinations of this word relate to the way that other (comparable) words combine. The extension has to do with the operator-argument niche that the given word already holds as compared with partially related words—words related not only in meaning, but also by the kinds of combinations that they commonly enter into.

Indeed, meaning alone does not suffice to determine how to extend the environment of a word. If we look at words that have moved into further environments, and so have taken on slightly different meanings, it will be clear that the meaning of the word before it entered the new environment is not sufficient to explain why it was extended into that environment. The meaning of *sail* before it was used for Zeppelins would not quite account for why *sail* came to be used for them (rather than *float, fly, soar,* etc.). The meaning of *fly* in its uses other than with *flag* does not readily explain why *fly* rather than various other words came to be used with *flag*. Therefore to go from meaning to the extension of environments of a word would not be very satisfactory. In contrast, we can go from the environment of a word to its meaning, but only from the meaning of that environment: if we know the meanings of the words with which a given word occurs, we can closely estimate the meaning of the given word. Indeed a dictionary that goes into detail on the meanings of words has to give examples of the use of the given word in respect to others. This relation among meanings is partly circular, but it shows the role of word combination in specifying meaning.

It is also relevant to consider what kind of meaning words have. The meaning of words in sentences is not like the meaning to a person of the objects and events in life (although certain words, such as one's own name, or the title of a cherished work of art, may be invested with the latter kind of meaning). Rather, the meaning of a word is, overtly, its association with certain objects or states in the world which are called its referents. However, few words have unique referents. The great bulk of words can refer to any of some set of objects or states or relations, including ones that they have not yet been used for. The meaning of a word might thus be better characterized by a sense, i.e., the properties common to its possible set of referents, than by a denotation, i.e., its extension or set of referents.

Notoriously, however, these properties cannot be specified sufficiently to constitute an adequate differentiation of word meanings. At this point, we can appeal to the availability of another possible differentiation of word meanings, namely the different operator sets or argument sets in respect to which a given word has selectional (higher than average) frequency. We can characterize the meaning of a word by these (not fully specified) sets, or by the composite meaning suggested by its relation to these sets. Then, for example, the problem of words with null extension, which has been much discussed in logic, becomes more malleable. Thus *unicorn* and *centaur* may be indistinguishable in the extension of their referents, since no objects exist to which either refers; but they have neither null nor indistinguishable meanings, since each occurs under a reasonably specifiable, and different, selection of operators.

3.3 Grammatical Meanings

One other kind of meaning appears in language, and that is grammatical meaning. There are a few specific meanings that are formed by the very event of combining, or of dependence, and that contribute to the meaning of the combination over and above the meanings of the participating words. The fundamental one is the meaning of the partial ordering, wherein the dependent word is understood as being "about" its arguments, the words on which it depends. In addition, the likelihood differences create the property of meaning differences among words; the meaning differences do not exist comparably in, e.g., mathematical notation, where likelihood differences don't exist. Another grammatical meaning is that of being a modifier, which in English is formed by bringing a secondary sentence as near as possible to the host, i.e., to the word to which it is secondary (and by the *wh* pronouning of the secondary repetition of the host: both in *John, who failed, must drop out* from *John—John failed—must drop out;* and in *He passed, which surprised me* from *He passed; his passing surprised me*). Yet another is the meaning of being the topic of a sentence, which a word obtains by being brought to the front of its sentence. There is also a semantic character to the status of grammatical ambiguity, which a sentential word sequence reaches by being derived from two or more different source sentences; and to the status of being nonce, when a word is used, artificially or jocularly, in a combination where it is not at selectional frequency.

Grammatical meanings, which are meanings of relations among words, can be named and described by words. But the sentences in which this is done will already contain the basic grammatical relations and their meanings.

3.4 Added Structure Adds Information Capacity

To have a picture of how the structural properties of words and sentences affect the information that these carry, let us now consider how increments to the constraints on word combination produce increments to the informational capability of a language system. This is just to give some examples of how little it takes, but what it takes, to create informational capacity.

The first example is the ability of words to refer to words. We begin by asking about possible ways of referring to objects and events. If one does this by pointing at them, or by drawing them, one would obtain a set of pointings or drawings that refer to, or indicate, or "mean" the objects or events in question. One could also refer to pointing or drawing itself, by pointing at it or drawing it. Each pointing or drawing is a token, not a type: we cannot distinguish in a principled way all pointings at a given object from all other pointings. Thus one could hardly distinguish diverse pointings by pointing at each separately, in a way that might establish a system of pointings. A different situation arises when objects and events are indicated, or referred to, by phoneme sequences, words. The phoneme sequences are themselves each a definite type, characterized by its phonemic distinctions, and their occurrences in sentences are ordered in respect to each other in specifiable ways. Therefore one can refer to words and to the occurrence of words in sentences, no less than one can refer to other specifiable objects and events in the world. Words can thus refer to words, though not to their own occurrence, as will be noted below.

The fact that a word is able to refer to words brings to language three major extensions of capability. One is that it makes possible a metalanguage—in particular a grammar—of language, as a subset of a language. This is the set of sentences in which the words refer to words, as seen in the preceding lecture. The second is the set of metasentential operators, that is, operators that operate on the occurrence, or the saying, of sentences. It consists of certain occurrences of words such as *say* on a sentence, with certain words that can be added to it (as further operators) as in *gainsay (deny)* or *ask*. The third is the apparatus of cross-references, i.e., of referentials and locating-pronouns that refer to nearby word occurrences.

First, consider the metasentence apparatus. For a person to say *I say yes* or *Yes is hereby said* means the same thing as "*Yes*" in quotation marks or as a *yes* that has undergone the activity of saying. Whereas if a person says *I said yes* or *He says yes*, it does not mean that *yes* has thereby been said; it merely means that the speaker claims the saying of *yes*. But if he says *I say yes* or "*Yes*," then this is not only a claim, he has thereby said it. In view of this, the *I say* contributes no information to the saying of the *yes* sentence and it is therefore zeroable by the conditions that were given in the first lecture.

There is much evidence of varied kinds that this *I say* can be assumed to have existed in principle on every said sentence. For instance, in every sentence one can make an interruption such as *to tell the truth* or *not to repeat myself*. Now, such phrases have to have been reduced from *for me* (perhaps in some cases *for somebody*) *to tell the truth* (1.6) and *for me not to repeat myself*. We know that the subject of the interruption is *I* because of the *myself*. For this *I* (or *me*) to have been zeroed in the interruption, it must have been zeroed as a repetition of a prior *I* that has since been zeroed. This prior *I* must have appeared with some verb, which has also been zeroed. There is strong reason for saying that this *I* plus verb was a "performative" *I say* on every said sentence, which is zeroable upon the saying of the sentence. Many grammatical constructions can be explained by assuming this. One is that the three set-theoretic operators, *not, and,* and *or,* which exist in language but differ grammatically from all other operators in language, can be derived from modifiers on this *I say*, thus explaining the peculiarities of their behavior in grammar. (For example, *and, or* are the only conjunctions which cannot take the form of a verb and cannot nominalize the sentences which they conjoin.) Another metasentence construction is tense. Tense is unique, not relatable to any otherwise known grammatical entity. Nor does it act overtly like an operator, which it should be, according to the theory that is being presented here. It can be shown that tense, in languages like English in any case, can be derived (reduced) from *before* and *after* as a conjunction between a sentence and the *I say* on that sentence.

A third kind of evidence for this *I say* is seen in certain words which can be best explained by deriving them from adverbs (modifiers) on the later-zeroed *I say*. One example is *any,* which means "every" in sentences such as *He'll eat anything,* but "(even) one" in *Will he eat anything? He won't eat anything.* We can obtain both meanings when *any* is derived (with historical justification) from *one* as adverb (*one by one*, in the sense of *for each*) on the *I say* and *I ask*, yielding roughly *I say, one thing by one* (or: *for each thing*) *that he'll eat it* (or: *He'll eat that thing*), reducible to *He'll eat anything*, and yielding *I ask, for each thing, whether he will eat it* (reduced to *Will he eat anything?*) and *I deny, for each thing, that he will eat it* (reduced to *He won't eat anything*). The two meanings of *any* come out naturally, one under *say*, the other under *ask, deny*.

Another example is in such words as *scarcely, hardly*. If I say *He is hardly the one to ask*, it is not clear what *hard* is doing there. However, one can derive this sentence from *I would hardly say he was the one to ask*, meaning "I would only with difficulty say that he was the one to ask." The connection of *hardly* to *hard* is understandable if *hardly* arose as a modifier of *say* and not of the sentence under *say*. So much for the metasentence operator, which results from the ability of words to refer to words.

The final construction that is due to this ability is cross-reference. It was seen in the first lecture that words can be pronouned or zeroed, when they refer to the same thing as another word occurrence at an easily located point in the sentence (1.5–6). In the relative clause, when the second occurrence of *John* was reduced to the pronoun *whom* (in *John whom I have long distrusted is responsible* from *John—John I have long distrusted—is responsible*), the first occurrence was immediately before the second—after the second *John* was moved to the front of its sentence and that whole sentence was used to interrupt the first one. Under *and*, the second occurrence, e.g., of *played*, is reduced to zero in *Casals played Bach but Schnabel mostly*

Beethoven from *Casals played Bach but Schnabel played mostly Beethoven;* the locating condition here is that the first *played* occupies the same position in the partial order of its component sentence that the second *played* occupies in its sentence. The conditions for reducing a given word involve the presence of the same word at a nearby position. How is the requisite information about sameness and location given? If we could do it in a metalanguage, we would simply state the conditions as they were stated here, and the nearness of the first occurrence of the word might not be so important as long as the location could be specified. But it has been seen that natural language has no external metalanguage, and that any metalinguistic information would have to be in a metalinguistic subset of the language itself or in another natural language—in either case in a language that already has such cross-reference, the problem then being how that language in turn got the information required for pronouning.

Language can give this information by adjoining to the given sentence, as an interruption right after the second occurrence of a word, a subsidiary metasentential sentence stating that another word occurrence, at an easily stated location in the given sentence, is the same word. On this information the second occurrence can be reduced to *whom*, zero, or whatever, and the subsidiary metasentential sentence is zeroed, having no longer any informational contribution. In this way something that would seem to be beyond the capability of language, something that should require an external and more powerful metalanguage, is performed in the language itself and within its rules, using only the additional power of words' ability to refer to words and to their occurrences. Because of this self-help or makeshift device, the location of first occurrence must, except in special circumstances, be in the same sentence as the second occurrence—this being the point at which the subsidiary sentence is adjoined—and the location is limited relative to the second occurrence (as being next, corresponding in partial order, etc.), rather than being in any position that can be identified absolutely, i.e., in terms of the sentence structure. We can see why only simple relative locations can support cross-reference.

It is thus not chance that the word occurrences that can be reduced to relative pronouns (*who, whom, which,* etc.) are precisely those that can be brought to the front of their interrupting sentence, where they come to be right after the first occurrences of the same word in the interrupted sentence (as in the two occurrences of *John* above). Words that cannot be brought next to each other, or that do not have a structurally easily specified relative location, are not cross-referenced: e.g., the second occurrence of *Mary* in *Mary came back and everybody greeted Mary*. (Note that we are speaking here only of cross-referencing, which is pronouning between two specified locations. Other pronouns, such as *he, she, it,* and *one* do not require a specified relation between the two locations of the repeating word, and indeed one cannot in general specify their antecedent on purely locational grounds. Their meaning to the hearer is "one mentioned nearby" rather than "the one mentioned at a given relative location.") Of course, all this is not to claim that such subsidiary sentences giving the cross-reference information were ever said, but that the amount of information needed to permit the cross-referencing that exists in languages is of this type, and that cross-referencing pronouns can be formulated in a system that contains no pronouning so long as there are words that can refer to word occurrences.

There are also other changes in structure that lead to changes in information capacity. One is the fact that under many conjunctions there is an increased likelihood that one or another of the words in their second component sentence should be the same as some word in their first component sentence (1.5). After, for example, *The farmers demand tariffs because* . . . the frequency of a repetition—the likelihood of one or another word of the first sentence appearing in the second—is greater than what would be due to chance. Furthermore, in many cases where there is no repetition (including pronouns), some of the words of the second sentence have a certain semantic relation to words of the first: a relation that can be expressed by intervening word-repeating sentences. For instance, if the continuation of the example above is *because imports are too cheap,* we can always find intervening sentences that supply the word repetition, e.g., *The farmers demand tariffs, which would raise the price of imports, because imports are too cheap* (note that *which* here is a pronoun for *tariffs*). This repetitional likelihood is not as great under many nonconjunctional operators on two sentences, e.g., in *A meeting of the officers preceded the reading of papers.* The greater expectation of repetition under certain bisentential operators that take conjunctional form gives them a more connective meaning than is felt in many of the other bisentential operators (such as *precede*).

Another structural addition that adds informational capacity is the added constraint of discourse, i.e., of sentence sequences that originate as connected stretches of speaking or writing. In discourses we find not merely the repetition of individual words that is seen around conjunctions. Rather, we find in each discourse some subsets of words that have a property special to that discourse, namely that a fixed operator-argument relation between words of these subsets repeats in many sentences of the discourse. For example, consider the following sentences (from an article by K. Lindestrøm-Lang and J. A. Schellmann in *Biochemica et Biophysica Acta* [1954], 15: 156–57):

> The optical rotary power of proteins is very sensitive to the experimental conditions under which it is measured, particularly the wavelength of light which is used.... The diversity of the factors which affect optical rotation is in many ways an advantage.... One of the authors (J.A.S.) has for some time been engaged in a study of the rotatory properties of several proteins, including the effects of temperature, wavelength, pH and the denaturation reaction.

It was found possible to set up a few classes of words or phrases, chiefly H (*protein,* etc.), R (*optical rotation,* etc.), W (*sensitive to,* etc.), K (*conditions,* etc.), such that the same partial order of these classes occurs in all of these sentences and in almost all of the others in the article. This makes possible a tabular representation of the material (including V for *have*, etc., C for connectives, and M for metascience material).

The sentences in the table are paraphrastic transforms of the sentences cited from the article, obtained by reductions and other transformations which do not change the partial order of words and the substantive information in sentences. For the structure of a discourse, it is not necessary to use the actual aligned transforms, such as are in the table; these are in many cases, as here, stylistically uncomfortable. Rather, it suffices to show, in a standard procedural analysis of each sentence of the article, that, say, H, R, W, K combinations recur, and that in each case H and R are coarguments under *have* or the like, and that the HR component sentence and K are coarguments under W.

We see here that a repetitional constraint is active in discourse, namely the repetition of various predicational relations (including modifiers) among relevant subsets of words. That this structure has an informational effect is clear from its practical applicability—it makes possible the tabulating of language information, and an inspection of the course of data presentation and of argumentation. But, more generally, the informational effect is that a discourse is not just an arbitrary sequence of sentences. It is not just a predication or a set of predications, not just a set of sentential word structures. Rather, it is a set of variations on a given word-class structure, on a sentence type like HRWK, consisting of various related cases of the same sentence type. It is this that gives the semantic effect of discussion and analysis, going beyond the simpler information that arbitrary independent sentences can carry.

Canonical Structure of a Discourse

C	H	V	R	W	K	M
	Proteins	have	optical rotatory power	very sensitive to	the experimental conditions	
wh-	"	"	"	measured under	"	
, particularly	"	"	"	very sensitive to	the wavelength of the light which is used	
	"	"	optical rotation	affected by	a diversity of factors	that . . . is in many ways an advantage
	several proteins	"	rotatory properties			One of the authors (J.A.S.) has for some time been engaged in a study of . . .
, wh-	"	"	"	including the effects of	temperature	
,	"	"	"	"	wavelength	
,	"	"	"	"	pH	
and	"	"	"	"	the denaturation reaction	

A final kind of structural change that leads to change in information capacity has already been seen in sublanguages. Especially in science sublanguages and above all in mathematics considered as a sublanguage, we see major differences in capability. For example, the restriction of science sublanguages to specified sentence types formed of particular word subsets excludes irrelevance from the system. And the elimination of likelihood differences within the partial order, together with the syntactic conditions for proof, enables mathematics to characterize structurally the kind of sentence sequence that precludes loss of truth-value between the initial sentences and the final.

3.5 Form and Content

The theory presented here, and the linguistic constructions that have been noted, enable us to make generalizations about the relation of form to content in language. It has been seen that the information carried by language is associated with, or expressed by, the constraints on equiprobability of words with respect to other words in the utterances, and this in a fixed way, step by step. The information is thus neither independent of this structure nor additional to it, but is an interpretation of it. Every occurrence of one of these particular constraints makes a fixed contribution to the information carried by the utterance, so that the information inheres in this structure. And the information of a sentence or a discourse is obtained in the course of structural analysis not because it was brought in at some point as a separate factor, but because it is yielded directly from the word choice and the stated constraints.

We also know why other features of language structure do not carry, or change, information. The phonemes do not carry a fixed meaning, or contribute in a regular way to that of their words, for if they did the universe of meanings, which must, at every time-slice, be preset (in the stock of word meanings) among the users of the language, would be limited by the limited stock of phonemes—all word-meanings would have to go back to regularities of combinings of phoneme meanings. The reductions cannot change meaning because they do not change the presence of the partially ordered words (even the zeroed words can be supplied by the hearer). The apparent shifts in the meaning of words are due either to the different operators or arguments with which they are occurring, or to the retained meaning of a zeroed operator or argument; this is also the case for metaphoric occurrences of words (*He peppered his speech with jokes* from a nonextant but structurally possible intermediate form *He treated-as-peppering his speech with jokes* from *He treated his speech with jokes as one's peppering something*).

This is not to say that every item of content that we may wish to express has a possible syntactic form. Feelings cannot be directly expressed in syntax. They can be talked about, using the ordinary grammatical structures; or they can be expressed in speech by nongrammatical devices such as nonsyntactic intonations (e.g., not the intonation of question, but the intonation of anger). Even if we consider information alone there are in language certain deficiencies in form-content correlation. Idioms are not a case of this. They are merely extreme cases of the general condition that the meaning of a word can differ when the word occurs with different operators or arguments. But there are cases of important deficiencies in what whole natural languages express or can express. One example is an informational capability that languages could in principal provide, but do not. This is the ability to specify, for each occurrence of a referential pronoun, what was its antecedent. Such an ability is available in principle because sentences are constructed, as we have seen, as partial orders of words, and are presented as a linear order of words. When the speaker says a pronoun, e.g., *she,* he knows which word occurrence this pronoun is referring to (i.e., of what word occurrence this *she* is a repetitional reduction). However, he is unable to communicate this, because, as we have seen, languages do not have the words and constructions that would provide an address for every position in the partial or linear orderings in their sentences. The hearer has to infer the antecedent of such a pronoun as best he can.

Another example is an informational capability that whole natural languages are in principle unable to provide. No natural language can exclude nonsense. It is impossible to define an effective grammar of a language in which all nonsensical sentences would turn out to be ungrammatical. The reason is that the first step in sentence formation is defined necessarily by the dependence-on-dependence relation, which says only that some words have zero probability in certain positions relative to other words. It cannot distinguish word subclasses, because of the lack of an external metalanguage in which to define them. The likelihood differences that distinguish meaning are defined upon this dependence construction, i.e., within the operator-argument relation. Although they may state that some words in the required class have vanishingly small likelihood under a given operator word, they cannot state that this likelihood is an unchangeable zero. That is, they cannot revise the results of the partial-ordering step—upon whose resultant they have been defined. The result is that sentences formed by the partial-ordering step can contain word combinations of very low likelihood, many of which may be nonsensical (since the main meaning of words is given by their selectional frequency under given operators). As has been noted, science sublanguages avoid irrelevance and nonsense by virtue of permitting (by definitions stated in their external metalanguage) only certain subclass combinations. In mathematics, the avoidance of nonsense results from having no likelihood gradations defined on the partially ordered sentential constructions: the zero-level words here are variables, which can take any value within their domain.

Yet another example of an informational capacity that is impossible in natural language is self-reference (more precisely, the self-referential use of cross-referential words). No word can refer to itself, to its own occurrence. In *This has four parts* the word *this* does not refer to itself.

The sentence says that we are talking about something (*this*) which is asserted to have four parts. *"This" has four parts* indeed says that the word *this* in that sentence has four letters; but its quotation marks (in speech, quotes intonation) are the trace of its having been reduced from *The word "this" has four parts* (from *The word that is "this" has four parts*), which is composed of *a word has four parts; the word is this* (where *this* is not referential). It follows that certain forms of the impredicative paradox do not exist in natural language. *This sentence is false* does not speak, by the grammar, of itself, but of another sentence which is referred to (cross-referentially) by the *this*. But the language provision seen here may not be proof against the paradox status of other forms, such as *The first sentence in this book is false* when that sentence is indeed the first one in the book in which it is printed, since here the *this* refers to something outside its own occurrence, namely to the whole book; it is overtly a demonstrative and not cross-referential.

Nevertheless, the great bulk of information in language can be structurally located—one might even say, is structurally based. This form-content relation may hold not only for the structure and information of sentences; there is good reason to think that it will be shown to apply to the structure of sequences or distinguished subsets of sentences. It has already been noted that science languages may have, or may be developing, certain "reasonable argumentation" conditions on sentence sequences, by which certain attested statements may be held to justify a related conclusion statement. Going beyond this, Henry

Hiz has long presented ways in which the estimated truth-status of sentences in natural language could be related to the set of their consequence sentences; many of these consequence sentences could be shown to be lexically or grammatically related to the given sentence—lexically by such relations as synonym and classifier, grammatically by reductions and other transformations.

3.6 The Structure of Information

We began with information as some kind of concomitant of language, undefined and of no known structure. Some kinds of information were found in language but not in grammar, such as alliterative allusion, and the intonations of anger and surprise, which do not combine in any regular way with grammatical constructions. Some were found in speech rather than in language, such as the individual voice, which gives information about the speaker. These aside, it was found that grammar, the set of structural regularities of language, was produced by certain constraints on the equiprobability of words. Within this, the constraints that had the greatest effect on these probabilities, the ones that determined which words co-occurred (i.e., were present together) in a regular way in the utterances, were precisely the ones that contributed most specifically to the information carried by the utterances. ("In a regular way" means primarily "in the same way in long utterances as in short.")

We therefore consider the information in the constructions that are made under just these constraints. Part of the information is clearly determined by the choice of words, which in the present formulation consists in their likelihood of co-occurrence. And indeed if we take any unordered set of words, say *John, Mary, Tom, Smith, call, tell, will, to, and,* we obtain a certain amount of joint meaning from the word collection alone. But if we hear *Smith will tell John to call Mary and Tom* or *Mary will call John Smith and tell Tom to,* or the like, we obtain much more specific information out of these words. Something has obviously been added here over and above the word choice, and this is the various grammatical relations, all of which can be derived ultimately from the partial ordering of dependence in word occurrences. Now, the word choice collection as above, i.e., the word likelihoods, is a constraint on the equiprobability of word co-occurrence—what can be called departure from randomness, a redundancy, in the set of word occurrences; it expresses the meanings of the individual words in their combinations. The partial order of dependence is also such a constraint, a redundancy; it expresses a meaning that may be called predication. Whether we think of the partial order imposing a redundancy on that of the word choice, as in the example above, or the word likelihoods imposing a redundancy on that of the partial order, as in the first lecture, in either case the information of the sentence was specified not just by the constraint inherent in a collection of meanings, but by a constraint acting on a constraint, that is, by a redundancy operating on a redundancy. A word by itself, e.g., *tell* or *book,* has meaning but does not in general provide specific information. So also for predication by itself, as a concept. But having one constraint (with its meaning)—e.g., the partial order—acting on another—e.g., word choice—yields something rather different, a specific sentential item of information.

There are also various further constraints on word likelihood, around conjunctions, in discourses, in science sublanguages, and in mathematics (as noted in 3.4). Each of these imposes additional redundancies and contributes characteristic information capabilities to the linguistic material involved.

In mathematical Information theory, developed within statistics, what was studied was the amount of information, as limited by the total amount of redundancy in a system. Information theory did not deal comprehensively with the relation between the individual contributions to the total reduncancy, and did not characterize the individual items of information that were involved in amassing the amount of information. The study of the structure of language not only shows its relation to information but also shows a structural characteristic of information itself, at least with respect to the information carried by language: that each item of information consists not only of a constraint, or a redundancy, or an isolated meaning, but of one constraint acting upon another.

3.7 What Is "Referring"?

In sum, the relation between language structure and language information becomes strong enough to be fruitfully analyzable only when the structure is described with maximal economy as a system of constraints on word occurrence. When this is done, it may be that we also reach the possibility of understanding the relation of referring itself. For if language structure and information are systems of particular departures from randomness, so are the objects and relations of the perceived world that language talks about. There is no basis here for any general claim that language mirrors the world it talks about, or that the structure of language and information corresponds to the structure of the world. However, we can see in the languages of science that their classes of entities and relations are distinguished vis-à-vis each other comparably to the entities and relations of the science itself. If the activities of the scientist in the laboratory differ completely from the activities of the objects of the science, and can apply to many events in the science, so do the metascience operators differ completely (in word classes and in partial-order position) from the operators and arguments of the science language proper; and the same metascience operators can apply to a great many science language operators. And if in immunological events objects that are antigens have a different standing than objects that are antibodies, with only particular classes of events holding between them, so in the immunology language there is a class of antigen words and a class of antibody words, with particular classes of operators or constructions occurring between them.

Although those correspondences are limited and flimsy, they suggest that the power of the sentences of language to refer to the events of the world, and more particularly the power of science language sentences to refer to the events of a science, involves some kind of similarity in the way that what is common to both—the underlying departures from randomness—is structured in a set of events, on one hand, and in a set of sentences, on the other.

4

The Nature of Language

4.1 The Structural Properties

THE SURVEY of language structure and information seen so far leads to certain conclusions about the nature and development of language. It was seen that language is a public activity, with the sound elements and word elements preset by convention within a community, in a manner that would not have been needed for private expression or even for interpersonal communication. This, together with the discreteness and repeatability of the elements (which reduces error compounding in transmission), and the lack of grammatical devices for direct expression of feeling, suggests that language developed primarily in the transmission of information within a public, rather than for personal or interpersonal use. In this connection it is relevant that the more efficiently the structure can be formulated along these lines, the sharper is the correlation between its thus-formulated structure and its information.

It should be recalled that we dealt first of all with simple, noncomposite words, whose phonemes are arbitrary with respect to the grammar and whose meaning is not sufficient to explain all of their behavior in the grammar. Hence, having the words alone is far from sufficient for the language. It was then seen that there was a dependence of the occurrence of words in sentences on certain other occurrences of words in sentences and that this dependence was a dependence on the dependence properties of the other words, not on any particular list of the other words. This dependence on dependence partitions the words into what we call operators and arguments. It was also seen that there were differences in the likelihood of operators with respect to their arguments and vice versa, and that the high-likelihood and low-information word occurrences could have their phonemes reduced, even to zero, without losing the words involved or the role that the meanings of the words play in the sentence. It was also noted that all events—the entry and reduction of words and the imposition of grammatical relations—are ordered in the making of a sentence. The result is that construction is contiguous: parts of a construction are contiguous to each other, and constructions as a whole are contiguous to related constructions, either as being next to them or as being nested within them. It was also noted that these constraints sufficed to analyze the sentences of the language to any detail desired, and that each step in the making of a sentence, each constraint, had a fixed contribution (substantive or subjective) to the information of the sentence.

The structure as formulated here has certain mathematical properties. To appreciate this, one should take into consideration that there are two kinds of applied mathematics. The usual kind is calculational, as in linear transformations or in the special functions of physics. Another kind is the finding of mathematical objects in real world situations. Such a finding is of interest if it is not only a naming of things, but also of some utility and some broad relevance to what the mathematical object does. In the case of language there is very little use of calculation. What there is, is the finding of a mathematical object, a system defined by its relations alone. As has been seen, the occurrences of words, or word sequences, in utterances of a language form a set of arbitrary objects, which is closed under the dependence-on-dependence relation, with each satisfaction of the relation constituting a sentence of the language.

What makes the words arbitrary objects is that none of their properties, other than their status in respect to the dependence-on-dependence relation, determines their behavior in sentences. The various compositions or representations—the phonemes, or letters, or Morse code, or single ideograms (as in Chinese writing)—serve primarily to identify which tokens (word occurrences) are members of the same type (words). Even this is not perfectly done, as in the case of different variants, spellings, or pronunciations of one word (e.g., *is-am-are,* or the two pronunciations of *economics*), or in two words that have the same spelling or pronunciation (e.g., *heart* and *hart*, or the three words *sound*). As to the meaning, it is a crucial property of words as separate entities. Even then there are problems, chiefly in that only for very few words can the meanings specify the precise range of referents. (Such specification is precise for the number words, but even the kinship terms have marginal cases.) Furthermore, in many cases the general meaning, or the meaning range, of a word does not quite apply when the word is in a sentence, i.e., is occurring in grammar, because the particular operator or arguments of the word indicate a particular meaning of it to the exclusion of the other meanings. To this extent, the meaning of a word in sentences is not quite identical with its prior meaning in isolation, when it is a separate sign rather than part of grammar. It follows from all of this that while the meaning of a word in sentences can be specified on the basis of its interword relations, and is closely related to its meaning when alone, its relation to other words is not adequately determined by this extragrammatical meaning.

Working then with just the relative occurrence of words in utterances, and seeking regularities that hold over all utterances, we investigate the relative frequencies or likelihoods of words, whence we come to the set of word sequences under the dependence-on-dependence relation. Starting with this set, we can define virtually all of the sets that are of importance for grammar—the elementary sentences, the unary and binary nonelementary sentences, the base sentences, and the reduced sentences. With the aid of all this, it is possible to construct or characterize all component sentences of a sentence, or all unambiguous readings of an ambiguous sentence, or, for example, a grammar common to any arbitrary two languages.

It is of interest to note what features are universal and what are not. The universal features in the structure of languages include: having phonemic distinctions, but not the particular ones that a language has; having words as phoneme sequences; having a linearized partial ordering of a dependence-on-dependence relation; and having likelihood differences among words, rather than all words having either random or identical or rapidly fluctuating likelihoods relative to the operators on them. Having reductions is also universal. However, all of these are only capacities.

There are other features in which languages are similar to each other to a large extent. The phonetic types are, for physiological reasons, rather similar. There are unusual sounds in certain languages, but many types of sounds are common. The number of phonemes is not very different among languages. Certain meanings are found in all or almost all languages. The main classes of words with respect to the dependence on dependence are much the same. Further, various types of reduction are rather similar in many languages.

The features that differ from language to language include the amount of grammatical complexity that is fashioned out of these constraints, which means how the constraints impinge on each other and where the complexities lie. There is total difference regarding: what phoneme sequences are used for what words, that is, for what meanings; the availability of morphology, which some languages have while others do not; and finally, just what reductions and what amount of reducibility the language has, and what kind of paradigms are thereby created in the grammar, including meaningful categories like tense and plurality.

4.2 Change

There is one other feature that is universal to languages, and that is that languages change. For certain purposes, it has been customary in modern linguistics to draw a line between the present and the past. A time slice—a moment in time—of a single language is important because many grammatical relations in a language hold only among coexistent forms in a language, those that are used by the same speakers: the forms have to relate to each other. But the line between the present and the past cannot be drawn precisely. In a sufficiently detailed grammar, there are always some forms that are in process of change. If a language is changing more or less all the time, in each time slice there are likely to be some forms or others that are not regular. One can regularize them, by assigning them to the nearest form from which they can be derived by some reasonable reduction. But in fact each language is not completely regular in detail. In certain grammatically exceptional forms one can see the earlier construction which has changed into the given form; in some cases the earlier is more regular than the form into which it has changed.

A few examples: There are certain verbs such as *give* and *take* which are used as pre-verbs attached to other verbs, as in *take a walk*. Originally they were operators on the following (e.g., *walk*), which were their argument. In their new use they have lost much of their meaning, as in *give a look at something* and *take a look at something*, which are not very different. Nevertheless, the grammar of these words in their original use has not entirely disappeared in their new use. Just as one can say, for the ordinary verb *give* (in today's speech and historically), both *give a book to somebody* and *give somebody a book*, one can still say *give it a look* as well as *give a look at it*. In the case of *take*, however, where one can say *take a book to the publisher* but less comfortably (and unhistorically) *take the publisher a book*, one can say *take a look at it* but not *take it a look*.

A different kind of example, which is revealing in its simplifiability, is seen in the auxiliary verbs *can*, *must*, etc. These differ from ordinary verbs in several respects: they do not take tenses (no present tense *cans*, no *will can*, no past tense unless *could* is taken as past of *can*, which is unsatisfactory in part); they cannot be nominalized (no participle and no infinitive—no *His canning to drive surprised me* from *He can drive*); the verb after them cannot have a separate subject of its own (no *I can him to leave* by the side of *I can leave*, compared with *I want him to leave* by the side of *I want to leave*). However, several centuries earlier these words had been tensed verbs, mostly in the preterite form which has since dropped out of English; hence it is understandable that they do not take a further tense, nor *-ing* (which is added only to the untensed verb, as in *his driving*).

Furthermore, there is some evidence that some of

these verbs had earlier permitted a subject on the verb following them, so that one could have said such a construction as *I can for him to do it* (the meaning of *can* having been "to know how"). Then the *can, must,* etc. words can be viewed as members, but with undetachable tense, of the class of operators like *hope, try, let*: operators whose first argument is a zero-level word (a concrete noun, e.g., *I*) and whose second argument is an operator (e.g., *drive, do,* carrying its arguments in turn, e.g., *he*), as in *I hope for him to win, I'll try for him to win, I let him win*. Their main other peculiarity is that the probability of the verb following them having the same subject as they themselves have has reached 1 (hence the second subject is zeroed). Thus the sameness and the zeroing of the second subject has become a requirement: *I can drive* is as though zeroed from a nonexistent (or no longer extant) *I can for me to drive,* to the exclusion of such sentences as *I can for him to drive.* Such a derivation of the auxiliary verbs locates them more closely in the interrelations of the other operators of the language, and presents their peculiarities as a matter of exceptionally high frequency of occurrence (for the lower subject to be the same as their own) and a matter of shape (the inability to detach the past tense; the dropping of the following *to* is found also in *I let him win*). This analysis makes the auxiliaries less irregular than if we consider them a unique kind of operator whose second argument is not a whole sentence, but just a verb without its required subject. (The present theory defines the set of sentences as the locus of words with requirements satisfied.)

Constructions such as in the above examples would

not be explained in a time slice grammar, but they are regularizable historically. On the one hand, in describing a particular grammar one has to take account of particular results of change. On the other, in a general theory of grammar one has to take account of the fact that there is change; the fact of change has to show up in an explanation of language structure as a whole. And of course, short of evidence to the contrary, one has to assume that change has been going on as long as language has been going on, not that it started at some particular moment. It might be helpful to note here that different parts of grammar change at different rates—and different languages change at different rates. What is reasonably likely to change is word use, the likelihoods of combination. Change is slower in the word stock, still slower in the grammatical constructions and the grammatical subclasses of words, and in the phonemic distinctions. What doesn't change at all, as far as we know about language, is the existence of a partial ordering. This property, which makes sentencehood, appears to be universal and invariable.

4.3 Stages of Development

Of the changes that go on in languages, the great bulk do not affect the structure of the language. They merely replace one word in certain environments by another, with various effects upon the interrelations of words. Some changes indirectly cause loss of structure, especially if they affect morphemes (words or affixes) whose occurrence is widespread or paradigmatically required. For example, the replacement of case endings (genitive, dative, etc.) by prepositions, which has taken place in many languages, does not alter the (partial-ordering) relation among the morphemes of a sentence (the preposition does the syntactic work of the case ending), but it can affect the grammar because the case endings constitute a small set that is required in certain positions, whereas the prepositions form a larger set which may occur without being required in some other situations although it is required in the positions where it replaced the case ending. There are yet other changes that create structure. For example, the definite article (*the*, in English) is not universal, and in languages in which it exists it has been formed in historic times by reduction, mostly from referential or "demonstrative" words such as *that* (as in *The plan is to deny all* reducible from *That which is a plan is to deny all*). The definite article enters into grammatical requirements which its less standardized source did not have.

So much for the changes in relatively recent times. If we consider longer periods, say the whole known history of English, or of French, we find recognizable structural differences piling up in attested ways. If we take the differences between any present Indo-European language and what is known of the common Proto-Indo-European language—a matter of just a few thousand years—we find much larger structural differences established by the comparative method, although the central framework of the language remains.

Aside from the structural developments due to

changes in language, we can also consider indications of development gleaned from the structure itself when the structure is described most compactly without what may be called "noise" in the description. That the structure may contain some evidence of development follows from the particular description that was presented in the first lecture. If the sentences of a language had been formed by various local types of combination, especially if these were determined simply by semantic compatibility, then the various words and their constructions could have arisen at much the same time or at scattered times. In contrast, the fact that all elementary sentences, needed for all other sentences, are formed by a single dependence relation which arose within word sequences means, as will be seen below, that unstructured word sequences are prior in description and could be prior in use to the development of this dependence within them. Furthermore, the fact that all sentences other than the elementary not only are derived from the elementary but specifically contain them means that elementary sentences are prior in description and could be prior in use to the formation of nonelementary sentences. It will be seen below that several developmental stages can be reconstructed for language from the nature of its constructions.

Consider first the arising of elementary sentences. In the analysis presented here, sentences are created by a partial order of words in respect to their combinability. This partial order does not create word combinations: it is not a concatenational operation that brings words together. Rather, it is a criterion characterizing those word combinations that are said. To see how the criterion could arise, we assume that many varied word combinations were said before this partial ordering developed. We can imagine that some words got to be used more and more with words of a particular informal set, because it made sense that way. For instance *eat* might be used with word pairs like *child, berry,* more than *eat* would be used alone, and more than it would be used with word pairs like *walk, sleep* (to say *walk eats sleep*). Then it is not hard to see that the occurrence of *eat* would come to be specially connected with the occurrence of such word pairs as *child, berry* to such extent that each occurrence of *eat* would be understood as associated with such pairs that are in use with it. So much so, that in their absence such pairs would be assumed, and if one says *eat,* one would understand such a pair to be implicit, i.e., that somebody is eating something. Under these standardizing circumstances, once one is accustomed to what pairs *eat* occurs with, not only are completely different word pairs such as *walk, sleep* not said with *eat* (to make *walk eats sleep*), but if they ever were said, it would be counter to expectation; it would in effect be "wrong." The arising of such standardization, to the point of institutionalization, is not unknown in social behavior even when no vested interests are promoting it. And once *eat* is not said except in association with a word pair of a particular set, such as *child, berry,* then when *eat* is indeed said with such a selected pair it has the effect of being said "about" them: it requires their class and it selects them in particular. This is approximately the meaning of this dependence, i.e., of the partial order. It is a meaning of saying something about or predicating about. Predication is an interpretation of the increasing standardization of this dependence.

With this development we have the elementary sentences of a language. As noted above, every other sentence, whether expanded by further operators or transformed by a reduction, contains one or more of these elementary sentences. For this to be the case, elementary sentences have to be defined before the nonelementary ones are, and can reasonably be thought to have existed before nonelementary sentences came to exist.

The move from dependence on a roughly given list of words or word pairs to the relation of dependence on dependence is a further development. The dependence of a particular word like *eat,* a first-level word, on particular other words, zero-level words like the pair *child, berry,* is a relation among words; it can be viewed as a relation among lists of words, of a word *eat* to a certain list of other words. The dependence of a second-level word, such as *continue,* on a first-level word, such as *eat,* as in *The child's eating berries continued,* is also a relation of a word (*continue*) to a word list. But, as we have seen, the relation of dependence is not that of a dependence on lists; it is the dependence on the dependence properties of words. The dependence on dependence is a generalization of the dependence of each word. And indeed, one does not define *eat,* or the whole set of operators on a pair of concrete nouns, by a list of arguments. Words enter and drop out of use as arguments, and too many words are borderline in their argument occurrences, so that speakers of the language may not be sure if they normally occur as arguments of a given operator. The argument set of *eat*, and of virtually every operator, has to be defined, for the speakers of the language, by a property rather than by a list; and this property is whether the argument in turn depends on anything or not—whether it is itself an operator or not. This property, which holds for arguments in elementary sentences as well as in nonelementary ones, could only have developed as a generalization, after the arguments existed in elementary and nonelementary sentences, informally defined by approximate lists. Only after the vaguenesses of the lists and the differences among them were in place could the generalization to dependence on dependence become the successful description.

Continuing the time ordering of sentence development, reduced sentences presuppose the existence of base, unreduced sentences. Even if a reduced sentence is constructed on the model, or more specifically the analogy, of other reduced sentences, it had to come after the existence of the base form of the reduced sentences that had served as model. In addition, the great bulk of reductions are optional, so that the source of the reduced sentence also exists (even if not always prior in time to the existence of the reduced form). The reductions bring many new features into language—grammatical ambiguities, fuzzy domains over which a reduction applies, sharp domains of reduction creating word subsets (e.g., the auxiliaries, the reciprocal verbs such as *meet*, affixes reduced from operators). These new constructions and features have to have come into existence later than the structure of the unreduced base sentences.

We see in all this several stages in the development of the set of sentences. We define here descriptive order to mean that A is descriptively prior to B if and only if the definition of B requires reference to A while the definition of A does not require reference to B. Descriptive order is partially ordered in respect to time order, in that if A is descriptively prior to B it can come into existence before B or simultaneously with B, but not later than B. In the syntactic theory presented in the first lecture, the dependence relation (of first-level word on zero-level arguments) was defined in the set of word sequences; nonelementary sentences (those containing a second-level operator) were defined in respect to elementary sentences; dependence on dependence was defined in repsect to simple (list-specified) dependence; finally, reduced sentences were defined in respect to base (unreduced) sentences.

In each case the prior construction could have existed for a long time before the development of the construction that arose from it. It must be understood that the relation of words to sentences (and so for the other stages above) is not like that of individuals to society: human individuals may not be prior in time to society, but words and unstructured word sequences could exist indefinitely long before people became habituated to the particular choices that standardized into the dependence. Even the neural equipment for a large stock of individual words largely in isolation may be so much earlier as to be independent of the neural equipment that processes the dependence relations between words as said in longer utterances. A major reason for thinking that long periods elapsed between stages is the fact that the processes leading to a next stage are in many cases not flip-flops but gradual institutionalizations of language use. This will be discussed below (4.4).

The question here is whether the descriptive order involves a time-ordered activity. Virtually nothing is known about how the brain processes the syntactic structure, nor is it to be assumed that the brain follows the whole descriptive order as against recognizing a large body of partly similar constructions. Nevertheless, some evidence comes from unintentional anticipatory slips of the tongue, as in Spoonerisms. In such slips, sounds in pronouncing an earlier word are drawn from a word later in the sentence. When they are unintentional they are mostly drawn from the argument of the given word (as in *queer old dean* from *dear old queen*, where *queen* is an argument of *dear*) or from a coargument of the given word (as in *Waste makes haste*, where the pair *haste, waste* are arguments under *makes* and coarguments of each other). In the partial order of dependence, the argument of a given word (operator) is prior to it, and should be known to the speaker—except in cases such as hesitations—before he says the operators; and the coargument is simultaneous with the given word and should be known to the speaker at the time the given word is said. In contrast, other slips, such as those drawn from the operator of the mispronounced word, are rare in unintentional slips, though—not surprisingly—not so in jokes: an example of drawing from the operator would be *Tathy can type* for *Kathy can type*.

Other developmental stages, which may have been interspersed with the above, are the alternative linearizations, both the fronting of words and the interrupting of a sentence by a subsidiary sentence. In the wake of these two, the grammar gains a major new construction: modifiers, both on words and on sentences. Another stage that presupposes the existence of its prior stages is the use of words to refer to words and their relative positions in an utterance, especially in the same utterance: the *say* operator, the referentials, and the metalinguistic sublanguage.

Then there are the constraints on sentence sequences, which presuppose the existence and structure of sentences. One is high repetition likelihood under certain conjunctions. Another is the fixed grammatical relation among discourse-specific word subsets which characterizes connected discourse. Both of these have been noted in the third lecture.

Finally, there are the massive new constraints in subject-matter sublanguages, which produce quite different grammars in the languages of science and in mathematics. Further, these sublanguages have a metalanguage external to them, located in the whole language. In the case of mathematics, we have not only the constraints of well-formedness on sentence construction but also the constraints on sentence sequence that characterize proof. Each of these properties could not have developed in the subject-matter use of language if the whole natural language had not been in existence first.

4.4 Processes of Language Development

How did these developmental stages come to be? In language change, it has been proposed, most convincingly by Manu Leumann, that syntactic change comes about by related sentences "competing" with each other for a communicational niche: of two sentences, both of which can be usefully said in given circumstances, one or the other may become more entrenched in use, to the point where one may finally replace the other in the language. In this way, change takes place in word use or in grammatical features (such as case endings on argument words, or on all except for some pronouns).

In surveying the partial order, the dependence on dependence, and the standardized reductions, we have seen that these fundamental processes of language can be understood as institutionalizations of customary use. The same can be said for specialized structures in various languages, which may be much later in the development of language. An example is the differentiation of operators into verbs, adjectives, and relational nouns (such as *father of*). In the theory presented here, we did not have verbs and adjectives and relational nouns; we just had operators (and their arguments). Consider three operators: *eat, close to,* and *father of*. In respect to the dependence relation, all of these require the same thing: two zero-level words, i.e., two concrete nouns. If we consider two other operators, *walk* and *ill*—that is, to be ill—we find that both require just one zero-level word. In many languages, all of these and similar operators are distinguished into verbs, adjectives, and relational nouns. They are distinguished on secondary grounds, on their relation to tense change. More exactly: the tenses can, arguably, be considered as though reduced from certain uses of *before* and *after*. Certain operators have relatively high likelihood of changing within the discourse, as between *before* and *after*. That is, they are less stative, and there is closer relation to differences in tense. These operators have the tense attached directly to them, and they constitute verbs. Verbs are thus those operators that have the tense attached directly to them, as having a greater likelihood of tense changes.

For those operators that do not have such a strong likelihood of tense change within the discourse, the tense is less closely involved with them. Here the tense is carried by a separate word, yielding adjectives, as in *was close to* in contrast with *walked*. As to those operators that were most stative, they took in addition the nounlike morphology of arguments (arguments being in general more stative than operators), and became relational nouns, as in *was a father of*, where *a* is a word attached to (countable) zero-level words, to concrete nouns. Note that this placing of the tense is not just a likelihood. It is not a custom but a demand; departure from it is an error. If a person changes intermittently from being ill to being well, one cannot for that reason treat the word *ill* in English as a verb in his case; one cannot say *He illed* just because the intent was more like a verb. *Ill* has to be treated statively even when it is not stative. Here we have use being codified into a rule, being institutionalized.

The institutionalization of use can be seen in language even in historical times, as in the formation of the periphrastic tenses in French and in English. The history of the perfect tense in French is clearer than in English, but the course of development is essentially the same, from freely formed sentences like a present-tense *He has the fish caught*, in the sense of "He has the fish in a caught state, in the state of his catching it" to *He has caught the fish*. In English, this last has the special meaning and time adverbs of a "perfect" tense (e.g., within this morning, not at 10 a.m.) and still carries various properties of the present tense; but in French the "perfect" form now bears the meaning and time adverbs of the past tense.

We have considered here the developmental processes in syntax. There are of course other processes in language. There is word borrowing from other languages. There is long-range phonetic change, which can lead to changes in the phonemic stock and can affect syntactic constructions (as in loss of case endings). Above all, there is analogic change, which is of first importance both in language structure and in language change. To a large extent the changes brought about by analogy and the other processes are reinterpreted in syntax as though they had been produced by known (or new) reductions acting on regular sentences that originally satisfied the dependence relation. Thus the effects of these other processes are to a large extent domesticated into the dependence and reduction system.

4.5 Language as an Evolving System

Certain general considerations about language emerge from this structural description. They suggest that language evolved and is evolving; we may still be at an early stage of it. That it evolved we see by the stages of development, and also by its accretional structure—for instance, that expanded sentences contain the elementary sentences, and that they are built from elementary sentences in the same way the elementary sentences are built from words. The expanded sentences are not a fresh start at sentencehood, nor is each derived from elementary sentences (i.e., obtainable in a regular way from them) in ways that nullify the structure of the original ones. Similarly, reduced sentences do not structurally supplant their source sentences; they merely reduce (or rarely permute) words in respect to their arguments or operators. In all of these cases the prior structure continues to exist, and an understandable additional relation acts on it. More generally, it is not merely of theory-construction interest that each new stage or construction is defined on prior ones, imposing a further redundancy on the prior redundancies; this property is crucial to an evolving system.

A few directions of development can be discerned in the stages and construction of this system. One is effectiveness in expressing information, or more specifically in transmitting it, seen for example in the discreteness of phonemic distinctions and in the presetting (i.e., learning) of words and their partial order (i.e., their operator-argument requirements). A second is the extension of informational capacity, which is seen in the use of phoneme sequences (words) rather than individual phonemes as minimal meaning-carriers, and in the openness of the vocabulary, and also in the growing informational capacity of the added constraints in the successive stages (3.4). A third is the stability and flexibility of the system, seen in the identifying of the argument domain of an operator by the dependence properties of the arguments rather than by fixed lists of them; seen in the public standardization of phonemic distinctions, vocabulary, argument requirement, and main reductions; and also seen in the ability to reinterpret the resultants of analogy as though they were derived in regular ways from (possibly nonexistent) source sentences. A final direction of development is economy—advanced in different respects and at different rates in various languages. This is seen in letting words have somewhat different meanings under different operators (or over different arguments), which reduces the stock of words needed to express the meanings in question, though at the cost of making meaning partly a word-pair property rather than a word property. It is seen most clearly in the reductions, whose individual compacting effect is more obvious than it is in the transformations composed of them.

The character of an evolving system is seen also in the

fact that language is not a perfected system, precisely suited to its use. This is seen in the fact that not all possibilities have been exploited: for example, both the partial order and its linearization permit the creation of an addressing system within language that would identify each word occurrence in an utterance, but no language has more than the barest rudiments of such facilities. It is also seen in the many inefficiencies that dot grammars at various points: in morphophonemics (given morphemes having different phonemes in different operator-argument situations); homonyms (different words having the same phonemes); ambiguity resulting from degeneracies in reductions; and irregularities (morphophonemic, borrowing, etc.) persevering in some of the most common words (e.g., in English *be*). Furthermore, some word-subclass properties do not follow through to make regular constructions on a regular word domain, but remain as just a bundle of properties differently affecting different words. An example is the English auxiliary verbs *can, must,* etc. Some words are partly in the set and partly not. In respect to omitting *to: You ought not go* but *You ought to go* (similarly for *need,* but note also *I needs must go*). There are also words lacking the other auxiliary properties but like auxiliaries in omitting *to: He let them go, He made them go.* In respect to not taking on tense or nominalization (*-ing*): *could, might, should* have some properties of being a past tense of *can, may, shall;* and for *need* we have *needed, will need, needing.*

There are also various informational distinctions which language structures touch upon but incompletely:

tense in nominalized sentences (as in *for him to have been there,* where the perfect tense has to be drawn upon to give the effect of the past); the comparative and the superlative forms; the use-mention distinction (including quotes, in writing).

One might also mention that languages seem to stop in developing a form as soon as an adequate apparatus is reached. Thus in many languages few operators have more than a two-part argument: *sleep* has one (*John slept*), *eat* has two, and in English a few words such as *put* have three (*John put the book on the table,* but no *John put the book*). The information carried by operators with three or four argument places (if such existed) can be given by constructions of two-argument operators acting on two-argument operators (e.g., *John placed the book so that the book was on the table*). Similarly, languages do not seem to have more than two levels of operator: operators on zero-level words, and operators at least some of whose arguments are themselves operators. Here too, higher-level operators, for instance ones whose arguments must be operators on operators, could be paraphrased by second-level operators acting on second-level operators.

In all of this one sees the character of an evolving system, not something fixed as an object.

There is another general property that throws light on the nature of language. Language is undoubtedly unique. But the individual processes that create language, as they were seen here, are not unique. The various constraints are not so entirely different from constraints that exist elsewhere in the world. Grammatical relation, such as being a subject or an object, is not something that is known anywhere else; but to have things depend on classes of other things and appear only if things of the other classes appear, is a kind of dependence that is not necessarily unique to language. Also, language is a demanding structure: some things are regarded as being in it and therefore right, and some things are not in it and are therefore wrong. But these demands, as we have just seen, can be understood as institutionalizations of a less demanding and a more naturally occurring use in the combinings of words. In other words, there are demands in language that are unique to language, but we have just seen that one can reach these demands by a process of institutionalization of custom, of convenience, of what makes sense. This does not mean that one can make language be simply whatever makes sense to the speaker. For language is public, an institutionalization. But it is important to know that the demands of language, the rules of grammar, are reachable as the end product of a process of useful insitutionalization, from something that is not demanding and not unique. And of course, the process of institutionalization itself is by no means unique, being widely known—for better or for worse—in culture and in social organization.

Just in order to see what there is in language, and whether it is unique, we can even in principle count the demands (the departures from randomness) in language. We can count the demands that suffice to enable a person to speak a given language. The reason we can count this is as follows: Each constraint as it has been described here deals with phonemic shapes, which themselves are combinations of phonemic distinctions that can be counted. One can count what distinctions are needed in order to establish the phonemes. Then each constraint deals with phonemic shapes and with the likelihoods of their occurrence with respect to each other. The likelihoods of their occurrence can also in principle be counted, by weighting the estimated relative frequencies in a few divisions—very rare, rare, selectional, very frequent. Thus, in each sentence the applications of the constraints are in principle countable. Furthermore, since each constraint is defined on the resultants of other constraints, we can arrange the counting activity to be sure that we are not counting anything twice, which is also very important. This means that in principle, it is possible to see just how much a person has to know in order to speak a language within a given vocabulary limit. Nobody will do this counting, but we can see that there is nothing magical about how much, and what, is needed in order to speak. Finally, and this is perhaps more important, we can see roughly what kind of mental capacity is involved in knowing each contribution to the structure—in knowing phonemic distinctions; in knowing the phonemic composition of words; in knowing the requirement status of words, i.e., their dependence on the occurrence of other words; in knowing the (mostly pairwise) likelihoods of operator-argument choice and the rough meanings attached to each word; and finally, in knowing the reductions in phonemic shape of given words in operator-argument situations. The kind of knowing that is needed here is not as unique as language seems to be, and not as ungraspable in amount.

The overall picture that we obtain is of a self-organizing system growing out of real life conditions in combining sound sequences. Indeed, it could hardly be otherwise, since there is no external metalanguage in which to define the structure, and no external agent to have created it.

Index

GPSR Authorized Representative: Easy Access System Europe, Mustamäe tee 50, 10621 Tallinn, Estonia, gpsr.requests@easproject.com

www.ingramcontent.com/pod-product-compliance
Lightning Source LLC
LaVergne TN
LVHW101438200926
848483LV00004B/397

* 9 7 8 0 2 3 1 0 6 6 6 2 4 *